ALSO BY HANS CHRISTIAN VON BAEYER

Taming the Atom
Rainbows, Snowflakes, and Quarks

The Fermi Solution

The Fermi Solution

Essays on Science

Hans Christian von Baeyer

Random House
New York

Library of Congress Cataloging-in-Publication Data
von Baeyer, Hans Christian
The Fermi solution : essays on science / Hans Christian von Baeyer.
p. cm.
ISBN 0-679-75570-5
1. Science. 2. Physics. 3. Physicists—Biography. I. Title.
Q158.5.V66 1992
500—dc20   92-56853

Manufactured in the United States of America on acid-free paper
9 8 7 6 5 4 3 2
First Paperback Edition

To H.J.v.B. and M.W.W.

Contents

The Fermi Solution

1.

The Fermi Solution

At twenty-nine minutes past five, on a Monday morning in July of 1945, the world's first atom bomb exploded in the desert sixty miles northwest of Alamogordo, New Mexico. Forty seconds later, the blast's shock wave reached the base camp, where a cluster of scientists stood in stunned contemplation of the historic spectacle. The first person to stir was the Italian-American physicist Enrico Fermi, who was on hand to witness the culmination of a project he had helped to initiate.

Before the bomb detonated, Fermi had torn a sheet of notebook paper into small bits. Then, as he felt the first quiver of the shock wave spreading outward through the still air, he released the shreds above his head. They fluttered down and away from the mushroom cloud growing on the horizon, land-ing about two and a half yards behind him. After a brief mental calculation, Fermi announced that the bomb's energy had been equivalent to that produced by ten thousand tons of TNT. Sophisticated instruments were also at the site, and analysis of their readings of the shock wave's velocity and pressure, an exercise that took several weeks to complete, confirmed Fermi's instant estimate.

The bomb-test team was impressed, but not surprised, by this brilliant bit of scientific improvisation. Enrico Fermi's genius was known throughout the world of physics. In 1938 he had won a Nobel Prize for his work on elementary particles, and four years later, in Chicago, had produced the first sustained nuclear chain reaction, thereby ushering in the age of atomic weapons and commercial nuclear power. No other physicist of his generation, and no one since, has been at once a masterly experimentalist and a leading theoretician. In miniature, the bits of paper and the analysis of their motion exemplified this unique combination of gifts.

Like all virtuosos, Fermi had a distinctive style. His approach to physics brooked no opposition; it simply never occurred to him that he might fail to find the solution to a problem. His scientific papers and books reveal a disdain for embellishment—a preference for the most direct, rather than the most intellectually elegant, route to an answer. When he reached the limits of his cleverness, Fermi completed a task by brute force.

To illustrate this approach, imagine that a physicist must determine the volume of an irregular object—say, Earth, which is slightly pear-shaped. He might feel stymied without some kind of formula, and there are several ways he could go about getting one. He could consult a mathematician, but finding one with enough knowledge and interest to be of help might be difficult. He could search through the mathematical literature, a time-consuming and probably fruitless exercise because the ideal shapes that interest mathematicians often do not match those of the irregular objects found in nature. Or he could set aside his own research in order to derive the formula from basic mathematical principles, but, of course, if he had wanted to devote his time to theoretical geometry, he wouldn't have become a physicist.

Alternatively, the physicist could do what Fermi would have done—compute the volume numerically. Instead of relying on a formula, he could mentally divide the planet into a large number of tiny cubes, each with a volume easily determined by multiplying the length times the width times the height, and then add together the answers to these more tractable problems. This method yields only an approximate solution, but it is sure to produce the desired result, which is what mattered to Fermi. With the introduction of computers after the Second World War and, later, of pocket calculators, numerical computation has become standard procedure in physics.

The technique of dividing difficult problems into small, manageable ones applies to many problems besides those amenable to numerical computation. Fermi excelled at this rough-and-ready modus operandi, and, to pass it on to his students, he developed a type of question that has become associated with his name. A Fermi problem has a characteristic profile: Upon first hearing it, one doesn't have even the remotest notion of what the answer might be, and one feels certain that too little information has been provided to find a solution. Yet, when the problem is broken down into subproblems, each one answerable without the help of experts or reference books, an estimate can be made, either mentally or on the back of an envelope, that comes remarkably close to the exact solution.

Suppose, for example, that one wants to determine Earth's circumference without looking it up. Everyone knows that New York and Los Angeles are separated by about three thousand miles and that the time difference between the two coasts is three hours. Three hours corresponds to one eighth of a day, and a day is the time it takes the planet to complete one revolution, so its circumference must be eight times three thousand, or twenty-four thousand miles—an answer that differs from the true value (at the equator, 24,902.45 miles) by less than 4 percent. In John Milton's words:

> so easy it seemed
> Once found, which yet unfound most
> would have thought
> Impossible.

Fermi problems might seem to resemble the brainteasers that appear among the back pages of airline magazines and other popular publications (Given three containers that hold eight, five, and three quarts, respectively, how do you measure out a single quart?), but the two genres differ significantly. The answer to a Fermi problem, in contrast to that of a brainteaser, cannot be verified by logical deduction alone and is always approximate. (To determine earth's circumference precisely, the planet must actually be measured.) Then, too, solving a Fermi problem requires a knowledge of facts not mentioned in the statement of the problem. (In contrast, the decanting puzzle contains all the information necessary for its solution.)

These differences mean that Fermi problems are more closely tied to the physical world than mathematical puzzles, which rarely have anything practical to offer physicists. By the same token, Fermi problems are reminiscent of the ordinary dilemmas that nonphysicists encounter every day of their lives. Indeed, Fermi problems and the way they are solved not only are essential to the practice of physics, but also teach a valuable lesson in the art of living.

How many piano tuners are there in Chicago? The whimsical nature of this question, the improbability that anyone knows the answer, and the fact that Fermi posed it to his classes at the University of Chicago have elevated it to the status of legend. There is no standard solution (that's exactly the point), but anyone can make assumptions that quickly lead to an approximate answer. Here is one way: If the population of metropolitan Chicago is three million, an average family consists of four people, and one third of all families own pianos, there are two hundred and fifty thousand pianos in the city. If every piano is tuned once every five years, fifty thousand pianos must be tuned each year. If a tuner can service four pianos a day, two hundred and fifty days a year, for a total of one thousand tunings a year, there must be about fifty piano tuners in the city. The answer is not exact; it could be as low as twenty-five or as high as a hundred. But, as the yellow pages of the telephone directory attest, it is definitely in the ballpark.

Fermi's intent was to show that although, at the outset, even the answer's order of magnitude is unknown, one can proceed on the basis of different assumptions and still arrive at estimates that fall within range of the answer. The reason is that, in any string of calculations, errors tend to cancel one another out. If someone assumes, for instance, that every sixth, rather than third, family owns a piano, they are just as likely to assume that pianos are tuned twice in five years, instead of once. It is as improbable that all of one's errors will be underestimates (or overestimates) as it is that all the throws in a series of coin tosses will be heads (or tails). The law of probabilities dictates that deviations from the correct assumptions will tend to compensate for one another, so the final results will converge toward the right number.

Of course, the Fermi problems that physicists face deal more often with atoms and molecules than with pianos. To answer them, one needs to commit to memory a few basic magnitudes, such as the approximate radius of a typical atom or the number of molecules in a thimbleful of water. Equipped with such facts, one can estimate, for example, the distance a car must travel before a layer of rubber about the thickness of a molecule is worn off the tread of its tires. It turns out that that much is removed with each revolution of the wheels, a reminder of the immensity of the number of atoms in a tire. (Assume that the tread is about a quarter-inch thick and that it wears off in forty thousand miles of driving. If a quarter inch is divided by the number of revolutions a typical wheel, six feet in circumference, makes in forty thousand miles, the answer is roughly a hundred millionth of an inch, or a molecular diameter.)

More momentous Fermi problems might concern energy policy (the number of solar cells required to produce a certain amount of electricity), environmental quality (the amount of acid rain caused annually by coal consumption in the United States), or military technology. A good example from the weapons field was proposed in 1981 by David Hafemeister, a physicist at the California Polytechnic State University: For what length of time would the beam from the most powerful laser have to be focused on the skin of an incoming missile to ignite the chemical explosives in the missile's nuclear warhead? The key point is that a beam of light, no matter how well focused, spreads out like an ocean wave entering the narrow opening of a harbor, a phenomenon called diffraction broadening. The formula that describes such spreading applies to all forms of waves, including light waves, so, at a typical satellite-to-missile distance of, perhaps, seven hundred miles, a laser's energy will become considerably attenuated. With some reasonable assumptions about the temperature at which explosive materials ignite (say, a thousand degrees Fahrenheit), the diameter of the mirror that focuses the laser beam (ten feet is about right), and the maximum available power of chemical lasers (a level of a million watts is conceivable), the answer turns out to be around ten minutes.

Trying to keep a laser aimed at a speeding missile at a distance of seven hundred miles for ten minutes is a task that greatly exceeds the capacity of existing technology. For one thing, the missile travels so rapidly that it would be impossible to keep it within the laser's range. For another, a laser beam must reflect back toward its source to verify that it is hitting its target, which would be comparable to aiming a flashlight at a small mirror carried by a running man at the opposite end of a football field in such a way that the light reflected from the mirror would shine back into one's eyes.

The solution of this Fermi problem depends on more facts than average people, or even average physicists, have at their fingertips, but for those who do have them in mind, the calculation takes only a few minutes, and produces a result that is no less accurate for being easy to perform. Therefore, Hafemeister's simple conclusion, which predated President Reagan's 1983 Star Wars speech, agrees roughly with the findings of the American Physical Society's 1987 report entitled *Science and Technology of Directed Energy Weapons*, which was the result of much more elaborate analysis. Prudent physicists— those who want to avoid false leads and dead ends—operate according to a long-standing principle: Never start a lengthy calculation until you know the range of values within which the answer is likely to fall (and, equally important, the range within which the answer is *un* likely to fall). They attack every problem as a Fermi problem, estimating the order of magnitude of the result before engaging in an investigation.

Physicists also use Fermi problems to communicate with one another. When they gather in university hallways, convention-center lobbies, or cozy restaurants to discuss a new experiment or theory, they often first survey the territory by staking out, in a numerical way, the perimeter of the problem at hand. Only the timid hang back, deferring to the experts in their midst. Those accustomed to tackling Fermi problems approach the subject as if it were their own, demonstrating their understanding by performing rough calculations. If the conversation turns to a new particle accelerator, for example, they will estimate the strength of the magnets it requires; if the subject is the structure of a novel crystal, they will calculate the spacing between its atoms. Everyone tries to arrive at the correct answer with the least amount of effort. It is this spirit of independence, which he himself possessed in ample measure, that Fermi sought to instill by posing his unconventional problems.

Questions about atom bombs, piano tuners, automobile tires, laser weapons, particle accelerators, and crystal structure have little in common. But the means by which they are answered is the same in every case, and can be applied with equal success to questions outside the realm of physics. Whether the problem concerns cooking, automobile repair, or personal relationships, there are two basic types of responses: the fainthearted turn to authority—to reference books, bosses, expert consultants, physicians, ministers— while the independent of mind delve into that private store of common sense and limited factual knowledge that everyone carries, make reasonable assumptions, and derive their own, admittedly approximate, solutions. Stripped transmissions and severe depressions usually require professional help but more mundane challenges—preparing chili from scratch, replacing a water pump, resolving a family quarrel—can often be sorted out with nothing more than logic, common sense, and patience.

The similarities between technical problems and human ones is explored in Robert M. Pirsig's 1974 book, *Zen and the Art of Motorcycle Maintenance,* in which the repair and up-keep of a machine serves as a metaphor for rationality itself. At one point the protagonist proposes to fix the slipping handle-bars of a friend's new BMW motorcycle, the pride of a half century of German mechanical craftsmanship, with a piece of an old beer can. Although the proposal happens to be techni-cally perfect (the aluminum is thin and flexible), the cycle's owner, a musician, cannot break his reliance on authority; since the idea did not come from a factory-trained mechanic, it does not deserve serious consideration. In the same way, certain observers would have been skeptical of Fermi's analysis, carried through with the aid of a handful of confetti, of a two-billion-dollar bomb test. Such an attitude demonstrates less, perhaps, about their knowledge of the problem than about their attitude toward life. As Pirsig put it, "The real cycle you're working on is a cycle called 'yourself.' "

Ultimately, the value of dealing with the problems of sci-ence, or those of everyday life, in the way Fermi did lies in the rewards one gains for making independent discoveries and inventions. It doesn't matter whether the discovery is as mo-mentous as the determination of the yield of an atom bomb or as insignificant as an estimate of the number of piano tuners in a Midwestern city. Looking up the answer, or letting some-one else find it, actually impoverishes one; it robs one of the pleasure and pride that accompany creativity and deprives one of an experience that, more than anything else in life, bolsters self-confidence. Self-confidence, in turn, is the essential prerequisite for solving Fermi problems. Thus, approaching personal dilemmas as Fermi problems can become, by a kind of chain reaction, a habit that enriches life.

2.

Creatures of the Deep

No one has sighted a yale in the wild. In fact, the animal has yet to be seen anywhere at all—though a yale, if one were to present itself, would be difficult to overlook. According to the thirteenth-century mapmaker and natural historian Gervase of Tilbury, the elusive creature has the body "of a horse, the jaws of a goat, the tail of an elephant, horns of a cubit [about two feet] in length, one of which can be reflected backwards as the other is presented forwards in attack, and which can move equally on water or on land." On his world map, Gervase granted possession of the yale to India, which was located somewhere in Asia, just below the Garden of Eden.

Faced with depicting uncharted regions of the earth, medieval cartographers, such as Gervase, populated distant continents and marginal seas with dragons and unicorns, as well as with more exotic inventions, such as the yale. The practice hints at the profound fear of emptiness that accompanies human encounters with the unknown. Dragons and unicorns, even yales, are less disturbing than the vast reaches of unexplored territory represented by blank space. Man feels driven to populate such vacuums, to domesticate the darkness.

Today, the blank regions on the map of human knowledge lie far beyond India and the antipodes, beyond the moon and the sun, in what is ominously called deep space. In exploring this territory, astronomers have come up against a disturbing enigma. They have uncovered evidence of the existence of a vast amount of invisible material whose nature is completely unknown. According to recent calculations, this strange stuff, called dark matter, accounts for most of the mass of the universe.

Speculation about the identity of dark matter has invoked everything from swarms of hypothetical elementary particles to various cold, dim stars—brown dwarfs—and such bizarre astronomical objects as cosmic strings. At last count, there were about two dozen candidates. To be sure, speculation is a necessary first step in any attempt to understand the unknown, but, normally, the facts severely limit the number of possible explanations and thus rein in the imagination. In this case, however, the empirical constraints are few—chiefly because deep space is inaccessible to experimentation. Thus, the imagination is less fettered than usual, and the truth more difficult to ascertain.

The problem of dark matter is more than a technical difficulty; it is a disgrace. Many physicists believe that we are on the verge of finding the ultimate description of the material world—the elusive theory of everything. But as long as most of what the universe is made of remains unseen, how can that ambition be realized? Dark matter also holds clues to how the cosmos got started and where it is headed. It is the key to understanding the very fate of the universe.

Eager to make headway on so portentous a problem, physicists have grown desperate, filling the blank spaces in their knowledge with twentieth-century versions of the yale. So, after four centuries of empirical science, fact has once again become intertwined with fiction in the description of the cosmos. It is as if, in lifting their eyes from the much measured Earth to the largely unplumbed heavens, scientists have also traveled back in time. The discovery of dark matter has prompted a return to the Dark Ages.

The idea that there might be more to the universe than meets the eye cropped up as early as 1783, when the English rector John Michell conceived of a cosmic object so massive that "all light emitted from [it] would be made to return to it, by its own power of gravity." Today, such an object is called a black hole, and counts as a candidate for dark matter. But in the eighteenth century there was no way of testing Michell's hypothesis, and it was largely forgotten. In fact, only during the past decade or so, with the confluence of three hitherto separate branches of science—astronomy, cosmology, and elementary-particle physics—has the problem of dark matter begun to attract widespread attention.

The astronomical evidence for the existence of dark matter comes primarily from the motion of galaxies. The pinwheel rotation of these star clusters is far too slow to be noticed by telescope, but there is an indirect way of detecting it. As the stars at one edge of a rotating galaxy recede from earth, their light is drawn out like a concertina, and their colors shift toward the red end of the spectrum. Conversely, the colors of stars at the galaxy's other edge, which is approaching earth, are shifted toward the blue. These changes act as a kind of cosmic radar trap to reveal the speeds and directions of travel of astronomical bodies. In this way, astronomers have learned that spiral galaxies wheel through the intergalactic emptiness like Frisbees—too much like Frisbees, as it turns out.

Large celestial objects can rotate in one of two ways. The first type of motion is exemplified by our own solar system, with its great central mass (the sun), and smaller masses (the planets) at increasing distances from it. Each planet moves at a lower speed than its interior neighbor. Mercury, the innermost, races around the sun in three months, while Pluto, the outermost, requires two hundred and fifty years to complete one circuit. The reason the periods differ so much, beyond the fact that one orbit is longer than the other, is that the sun's gravitational attraction weakens with distance, and thus pulls each succeeding planet around more slowly. The second type of rotation is that of a Frisbee, which is characterized by a constant angular speed. This means that all parts of the disk, from hub to rim, take exactly the same amount of time to complete a revolution. The force required to sustain this motion is provided by the internal molecular attractions that hold together the plastic material of the Frisbee.

Since a galaxy consists of a central core of billions of stars, surrounded by increasingly nebulous spiral arms, it should revolve in the manner of the solar system—rapidly in the middle and slowly at the rim. But it actually moves more like a Frisbee, suggesting that some additional force is being brought to bear in order to give the spiral arms their unexpectedly high speed. That force cannot be anything but gravity. (Electricity, the only other force that works over galactic distances, is ruled out, because stars and planets are electrically neutral.) Since visible stars and clouds of gas account for only a fraction of the mass and, therefore, only a fraction of the gravity required to keep everything spinning at the observed speeds, a galaxy must contain reservoirs of undetected material, or dark matter.

Observations of the motions of stars in the Milky Way suggest that about half its mass exists in the form of dark matter. In the six most meticulously studied rotating spiral galaxies outside the Milky Way—deep-space objects with names such as UGC2885 and NGC3198—dark matter represents between 75 and 80 percent of the total mass. And judging from the motions of galaxies that make up galactic clusters, even higher proportions of dark matter are found in those systems. Having assembled these clues, astronomers estimate that an astounding 90 percent of the universe consists of this mysterious substance.

Cosmologists, for whom questions of what the universe was and what it will be are as important as what it is today, have approached the enigma of dark matter from the theoretical side. Einstein's theory of gravity, formulated in 1916, implies that the universe can have one of three possible structures and, correspondingly, one of three fates. Cosmologists agree that the universe burst forth some fifteen billion years ago from an infinitesimal but extremely dense kernel—an event known as the Big Bang—and has been expanding ever since. If the amount of matter and, therefore, the strength of gravity is sufficiently small, this expansion will continue forever and the universe is called open. If, on the other hand, the amount of matter is large, gravity will eventually halt and reverse the expansion and, billions of years from now, pull the contents of the universe back together in a cataclysm called the "big crunch." Such a universe is called closed. The dividing line between these two scenarios, in which the universe is neither open nor closed but "flat," corresponds to an amount of matter that is termed the critical mass. A flat universe will expand forever, but at an ever-diminishing rate.

Cosmologists estimate that visible matter represents about 1 percent of the universe's critical mass, which they consider astonishing because in their way of viewing things, 1 percent is practically as much as 100 percent. To most of us, of course, this seems patently wrongheaded. But cosmologists inhabit a world of numerical extremes. The number of galaxies in the universe, the number of stars in a galaxy, and the number of atoms in a star each contain so many zeros that a couple don't matter; a factor of a hundred is small potatoes in cosmology. Thus, from the cosmologist's standpoint, the fact that the universe contains just a hundred times less, and not, say, a billion times more or a trillion times less, than the amount of matter needed to make it flat is too remarkable to be a coincidence. They believe that the universe really is flat, that the amount of matter in it is, in fact, equal to the critical mass, and that astronomers will eventually find the 99 percent that's missing, if only they keep looking. For this reason, cosmologists welcome the evidence for additional matter that astronomers have found in the motions of stars and galaxies.

At this stage, particle physicists enter the debate. To account for the numerous constituents of matter within a unified framework, they have produced a hierarchy of theories, each accommodating more experimental data than the one that preceded it. Occasionally theoretical simplicity and elegance have required the introduction of new, seemingly fantastic particles, such as the fractionally charged quarks that protons are made of. But the strategy has been successful: paradoxes have been resolved, old theories streamlined, and, in many cases, the predicted particles found.

Of course the new theories haven't always worked out smoothly (for example, the concept of quarks had to be revised sharply when the particles initially failed to show up). But the general approach has paid off often enough so that physicists long ago shed whatever diffidence they might have had about proposing novel particles. When they became aware of the problem of dark matter, they knew just what to do. They dipped into their store of hypothetical particles to account for the missing mass.

The proposed candidates include theoretical particles called plancktons and magnetic monopoles; quark nuggets, electroweak nuggets, and strange matter; sneutrinos and heavy neutrinos; photinos, higgsinos, gravitinos, neutralinos, and magninos; axions and cosmions; polonyions and Weyl vector mesons. The properties of these particles are as diverse as their names are quixotic. Some are heavier than the heaviest known molecules; others are lighter than anything ever weighed. Some—the magnetic monopoles, for instance—rest on sound theoretical foundations and are being pursued by large groups of physicists the world over; others are merely straws in the wind. Certain kinds of heavy neutrinos are close to being eliminated from the list, whereas plancktons and polonyions are so elusive that we may never be able to confirm or rule out their existence. The only feature these particles have in common is that not one of them has been seen.

Appearances to the contrary, membership in the dark-matter club is highly restricted. The prospective particle must fit into the prevailing theories of matter and the evolution of the universe. It must interact with known particles, but only feebly; otherwise, it would have been detected by now. (The acronym WIMP, for weakly interacting massive particle, alludes to a class of such candidates' interactions.) Most important, the particle must exist in such unimaginably large numbers that it constitutes ninety-nine hundredths of all matter in the universe.

Of the imaginary particles that meet these requirements, the eeriest by far goes by the collective name shadow matter. The sole function of this mysterious stuff is to act as a source of gravitational force. It does not interact with ordinary matter in any other way. Unseen and unfelt, except for a gentle gravitational tug, shadow matter travels undetected through solid matter like a ghost. For all we know, it is passing through our bodies at this very moment. And there's the rub: For all we know, a herd of Gervase yales is also passing through our backyards at this very moment. In the realm of pure ideas, anything is possible.

If the quest to fathom dark matter has revealed anything, it is that speculation which breeds only more speculation, rather than inspiring new experiments and observations, does little to advance the cause of science. It may be a long time before physicists know what's really out there, and the answer may turn out to be much stranger than anything proposed so far. But meanwhile, instead of losing themselves in scholarly disputation of what might be, they must face the fact that there are things they don't understand. Science, in clearing away the fog of myth and mysticism that shrouded the world in the Dark Ages, has exposed not only sharply delineated islands of knowledge but also boundless seas of ignorance.

Like the imaginary creatures that inhabit the nether regions of medieval maps, shadow matter, as well as photinos, higgsinos, and most of the other dark-matter particles, performs a function, but it is one that has little to do with science. It makes a mysterious force a little less intimidating; it makes a largely indifferent and intellectually forbidding universe a little more comfortable. But it does not explain a thing.

3.

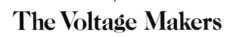

The Voltage Makers

With its shuttered villas and placid turquoise lake, the Swiss town of Lugano, near the Italian border, seems an idyll of tranquillity. But often, without warning, northern winds chilled by alpine glaciers sweep down to meet masses of warm air that drift up the Po Valley from the Mediterranean, producing thunderstorms with a violence unmatched in Europe. It was this meteorological quirk, not the splendid view, that brought the young German physicist Kurt Urban from Berlin to the summit of Mount Generoso, above Lake Lugano, in the spring of 1927. His aim was to harness the energy of storm clouds to drive the first atomic accelerator.

The inspiration for Urban's scheme came from Ernest Rutherford, the discoverer of the atomic nucleus, who had delivered a speech on the future of subatomic research to the Royal Society of London. To probe the atom further, Rutherford had said, it would be necessary to penetrate nuclei with electrically charged particles. Rutherford calculated that the particles would have to be propelled by at least a million volts, and he challenged physicists to find ways of achieving such power.

Urban's idea was to tame lightning, which involves more voltage than any other natural phenomenon on earth. Across a gorge on Mount Generoso, he and his team strung a two-thousand-foot cable. The free end of a grounded wire was then hoisted by pulley to a point sixty feet from the cable. Soon the scientists, huddled in a tiny observation shed, were rewarded by the spectacle of sparks leaping between the cable and the grounded wire. Urban estimated the electrical potential responsible for this display at eight million volts, a world record. But the experiment ended before a single atomic nucleus could be penetrated. On August 20, 1928, while adjusting the apparatus, Urban fell to his death.

Though the accident aborted the project, the quest for high voltage had begun. Even as Urban was preparing his ill-fated expedition to Mount Generoso, physicists in America were trying to produce lightning artificially in the laboratory. The most successful of these early attempts was based on the electrostatic generator, a device that produces high electrical potential, or voltage, by friction, the way clothing rubbed against the plastic cover of a car seat produces uncomfortable sparks in the dry air of winter. Electrostatic accelerators proved highly effective and soon surpassed Urban's record.

But then the race for high voltage encountered a fundamental hurdle: the same physical laws that govern the production of lightning also limit its artificial generation. When an electrostatic accelerator reaches a sufficiently high voltage, no insulator can prevent the onset of arcing—accidental lightninglike discharges from the apparatus to the ground. For this reason, electrostatic acceleration cannot exceed twenty-five million or so volts, an insufficient level for probing the innermost regions of the atomic nucleus.

When technology reaches a limit set by physical laws, there is no point in trying to bulldoze past it; progress can be achieved only by starting afresh along a different path. This was the case with electrostatic accelerators, as well as with later accelerators developed in the pursuit of higher voltage. By periodically revising their technique, physicists were able to design accelerators capable of generating millions, then billions, of volts. And progress still continues: the most ambitious machine yet conceived, the Superconducting Super Collider, or SSC, if constructed will produce twenty *trillion* volts.

The scientific allure of the SSC is easy to appreciate. In the machine, a stream of protons—positively charged nuclei of the hydrogen atom—will be compressed into a needle-thin beam and, guided by thousands of powerful magnets, shot into an identical beam approaching from the opposite direction. The crash will create miniature fireballs that fleetingly reproduce conditions of temperature and pressure similar to those that prevailed in the universe a fraction of a second after the Big Bang. The fragments of such explosions are clues to the origin and structure of matter.

But at the same time that the Super Collider introduces a new era in particle physics, the technology on which it is based threatens to bring the quest for high voltage to an end. Billed as the world's largest scientific instrument, the machine will be housed in a fifty-three-mile-long circular tunnel, and will cost several billion dollars to build. Given the limited funds available for scientific research, an even larger successor to the SSC, built upon a similar design, would be simply unaffordable. In that case, accelerators would go the way of the mammoth—

driven to extinction by the enormity of their own proportions.

To make matters worse, the SSC is not even the only machine of its scale. The European Center for Nuclear Research (CERN) has proposed its own accelerator, called LHC, for large hadron collider. (Hadrons, whose name derives from the Greek word for heavy, constitute a large family of particles that includes protons and other atomic nuclei.) Although the two machines differ in design, and in their maximum attainable energies, even the most enthusiastic advocates of big science find themselves hard-pressed to recommend the construction of two enormous accelerators that resemble each other as closely as do the LHC and the Super Collider.

For the first time in the history of subatomic research the obstacles to producing higher voltage lie outside the province of physical laws. Regardless of the scientific contributions the Super Collider and the LHC promise to make, they are surely the last of their species. If the search for the ultimate nature of matter is to continue into the twenty-first century, physicists must find other ways of producing high voltage. The time has come to chart a new route up the mountain.

In fact, the next era of accelerator physics may have already begun. A team under the leadership of Andrew M. Sessler of the University of California at Berkeley has proposed building a machine called the relativistic klystron two-beam accelerator that is considerably smaller and cheaper than its predecessors. The RK/TBA accelerates electrons, which weigh almost two thousand times less than protons and therefore move much more rapidly, so the machine's performance and characteristics cannot be compared directly with those of the Super Collider or the LHC. Nevertheless, the ultimate purpose of the device is the same as that of its larger cousins: to bombard nuclei with the fastest possible projectiles.

The RK/TBA is based on the idea that set the quest for high voltage on its proper course in the first place: resonance. In 1928, the year in which Urban died on Mount Generoso, Rolf Wideröe, a Norwegian electrical engineer at the Technical University of Aachen, Germany, decided to accelerate particles in stages, each one well below the limit set by arcing, rather than in one giant step. His accelerator consisted of a small vacuum tube into which copper rings, each attached to an external power supply, had been inserted at regular intervals. The first ring carried a negative charge that pulled positively charged atoms forward into the tube. As soon as the atoms had passed through the first ring, its electrical charge changed to positive, and the resultant repulsion pushed the atoms farther ahead. A little farther along the path the second ring then became negative to start the cycle of acceleration all over again. In this way, the atoms' speed was increased step by step.

Wideröe's pull-push scheme depended on the arrival of electrical impulses at the rings in strict rhythm with the motion of the atoms. Only the atoms that passed through the rings at the right moments—those that *resonated* with the applied voltage—were accelerated. To illustrate the idea, imagine a level trough in which marbles are propelled at regular intervals by paddle wheels. If a marble arrives under a wheel at the right time, it is kicked forward. If it arrives a bit early or late, it misses the paddle, and falls behind. Although the speed of the paddles remains constant, the stream of marbles accelerates as it travels along the trough, in the same way that successive small pushes timed in harmony with the motion of a swing can propel a child to great heights. Relying on this simple principle, Wideröe's miniature accelerator produced forty thousand volts.

As the limit to electrostatic acceleration was approached, resonance acceleration came into its own. The machines grew larger and more powerful, their design more sophisticated. Wideröe's rings were replaced by hollow chambers, called resonators, which are pierced by two ports for allowing the particle beam to enter and leave. The resonators receive their power through pipes transporting microwaves, instead of wires carrying electric currents. The microwaves, in turn, are generated by radio tubes called klystrons, from the Greek verb for the breaking of ocean waves.

A klystron is a cylinder in which a current of electrons streams from one end to the other, developing ripples along the way. Just before reaching the end of the tube, the ripples grow so large that they break, like waves breaking on a beach, and give up their energy in the form of microwaves. A modern electron accelerator is nothing more than a long string of resonators, each one fed microwaves by its own klystron, which, in turn, is powered by current from the commercial electrical grid.

The novelty of Sessler's design is that it replaces the standard array of individual klystrons with one single, continuous klystron. The entire machine takes the form of two long, straight, evacuated tubes joined at regular intervals by short copper pipes. The first tube is the giant klystron and contains a copious, pulsating stream of relatively slow electrons, while the second one acts as a connecting thread for a series of resonators that accelerate a much thinner stream of electrons to high speed. Microwaves continuously carry energy from the first tube into the second. The efficiency of this arrangement derives from the fact that it manages to capture the kinetic energy that is wasted whenever individual electron beams crash into the ends of the hundreds of separate klystrons of conventional accelerators.

The world's most powerful existing electron accelerator, the

Large Electron Collider at CERN near Geneva, delivers two sixty-billion-volt beams in a seventeen-mile circular tunnel. In contrast, if two straight RK/TBAs were aimed at each other, the entire assembly would develop six times more voltage, but its length would measure a mere two miles. Sessler's machine packs the strength of a mammoth into the body of a mouse.

The design of the RK/TBA suggests that it is the prototype for a new species of accelerator for which the futuristic name "beam transformer" would be appropriate. Ordinary transformers, which are essential components of all electrical and electronic technologies, convert voltage from high to low (from 120 volts to six, for my cassette recorder) or from low to high (from 120 volts to 220, for my French coffee maker). They do not so much generate or consume energy as simply trade current for voltage, or quantity of electricity for intensity, and vice versa.

A low-to-high transformer consists of two wire coils—the primary coil, in which voltage remains at a low level, and the secondary, in which high voltage develops. Energy flows from the primary, which acts as an oscillating electromagnet, into the secondary, where electrons are accelerated by the alternating magnetic force. To imagine a beam transformer, one has only to substitute electron beams in vacuum tubes for currents flowing through wires, and microwaves for the fluctuating magnetic field.

Besides spawning a new line of accelerators, the RK/TBA may help physicists resolve the current impasse in accelerator development. Because the factors affecting the Super Collider's construction are economic rather than scientific, the direction of nuclear and particle physics is now in danger of being decided largely by politicians and public-policy makers. With more than four thousand construction jobs, almost three thousand permanent jobs, and billions of dollars at stake, it's not hard to see why the scientific aspects of the project may play a secondary role in the decision-making process.

Against this background, it is reassuring to learn that less costly and more efficient ways of accelerating atomic particles are being pursued. The invention of the beam transformer demonstrates that more than half a century after Kurt Urban climbed Mount Generoso to capture the energy of lightning bolts, new routes to high voltage can still be found. Furthermore, it increases the likelihood that physicists will retain control over the quest to discover the innermost secrets of matter.

4.

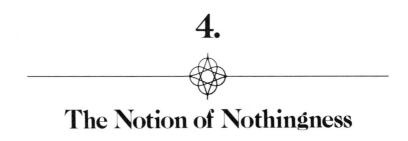

The Notion of Nothingness

According to an ancient Hindu cosmogony, the universe began with the all-pervading resonance of the magic syllable *om*. From this disembodied rumble the world came forth, and it will last as long as the sound reverberates. The appeal of the story, I suppose, lies partly in the gentleness of the process, compared with the violence of the Big Bang, and partly in our ability to reproduce the original sound ourselves, to reenact the Creation whenever the spirit moves us. There is something eminently humane, too, about starting with a word; indeed, the idea so suits mankind's sense of rightness that it is occasionally echoed in Western religious tradition as well. "In the beginning was the Word, and the Word was God," according to the Gospel of Saint John, and the great, dark emptiness trembled with the promise of incarnation.

Or did it? As understood by physicists, sound consists of waves of atomic vibrations that propagate through such materials as air or water or stone. No trembling of any kind can occur in the absence of a medium—something to do the trembling.

So matter either predated the primal invocation or appeared simultaneously with it; either way, the story of Creation out of pure sound doesn't hold up to scientific scrutiny.

Of course, it was never meant to. Trying to reconcile modern physics with four-thousand-year-old spiritual symbolism not only is fruitless but perverts the intention of both endeavors—science, to shed light on how the world works, and religion, to help us understand what the world means.

Nevertheless, at the core of this ancient story, and my half-serious critique, is an important scientific puzzle. For the idea of nothingness that underpins the earliest cosmogonies is just as baffling in the contemporary guise of the vacuum as it was in dim antiquity. We seem incapable of talking about the vacuum for very long without ascribing qualities to it or discovering things in it. Whenever a cogent definition is formulated, as happens from time to time, evidence turns up that requires physicists to fill the vacuum with complicated conceptual baggage. Then that baggage is thrown out, and the process starts afresh. So, instead of evolving, the concept goes in circles.

As recently as 1986 a paper published in the *Physical Review Letters* signaled yet another turn in this tortuous road: the discovery, according to the authors, of a *new* vacuum state. But how can there be different kinds of vacuums? Isn't this as absurd as saying that a coffee cup can achieve different states of emptiness? Isn't nothing nothing?

As long as science and philosophy were intertwined, nothingness was solely a theoretical concern. Parmenides, who founded the Eleatic school of philosophy around 500 B.C., and, with it, Western rationalism, believed that the universe, along with everything in it, was part of a single eternal reality he called *being.* Like the air that remains after the last note of a Brandenburg concerto has faded, being is something that can be thought about and described. In this view, *nonbeing,* or nothingness, is neither thinkable nor describable, and so Parmenides rejected it. Others, including his pupil Leucippus, took the opposing position that atoms, which compose matter, can move only if there is empty space—a void—for them to move into. And thus began a conceptual debate that continues among philosophers to this day.

When, in the early 1600s, experimentation joined logical analysis to launch modern science, nothingness came to be studied empirically. This early research relied on two devices. The first was the barometer, a thin glass tube sealed at its lower end and filled with mercury, then inverted with the open end immersed in a dish of mercury. When this was done, the column of silvery fluid that filled the tube dropped a few inches, leaving an empty space. Since no air could penetrate this space, what was left was a vacuum—sprinkled with no more than a few atoms of mercury vapor. A more versatile way of creating a void was made possible in 1650 by the invention of the air pump, which resembled a huge hypodermic syringe for sucking air out of closed vessels.

Thus the ancient question of whether nothingness exists was answered in the affirmative. Barometers and air pumps brought about the absence of air, or, more specifically, the absence of atoms and molecules. Although the creation of an absolutely pure vacuum poses technical problems that have never been fully overcome (even in the cleverest apparatus, and in the remotest regions of the universe, a few stray atoms are always present), one can nonetheless imagine, without running into a philosophical cul-de-sac, a space entirely bereft of atoms.

But is this space really nothingness? Is it possible that even after all the atoms are removed from a vacuum something may still be left? And what about space on a smaller scale? What occupies the interstices between the atoms of solid matter? Or even more intriguing: What's *inside* an atom, between the outer shell of electrons and the nucleus? Questions such as these arose when physicists began to suspect that despite the absence of any form of known matter, a vacuum may not be empty—an odd and unexpected piece of speculation prompted by discoveries about the nature of light.

In the latter part of the nineteenth century all waves were thought to require a medium: water for ocean waves, stalks of grain for undulating amber fields, air or bone or metal for reverberations of sound. Light, too, consists of waves, as James Clerk Maxwell proved in his unifying theory of electricity and magnetism, and so it should require some substrate for its existence. Instead, however, it displays the puzzling ability to traverse apparently empty spaces, devoid of atoms. Light from the stars reaches us through a vast celestial vacuum. Or, to use a more homely illustration: when air is removed from a glass jar containing a bell and a flashlight, the sound of the bell disappears, yet light continues to shine out.

This paradox weighed heavily on the minds of physicists around the turn of the century, and eventually drove them to invent a medium for the propagation of light, called the ether, which was supposed to fill all space, down to the gaps between atoms and even the emptiness within them. The ether filled the vacuum; where once there had been nothing, now there was something.

But, as would often happen in the course of rethinking nothingness, the introduction of this shadowy something into the otherwise pristine emptiness of the vacuum produced a contradiction. In general, the speed of waves increases with the stiffness of the material through which they migrate. Sound, for example, travels much faster through steel than through air, because the atoms of a stiff substance are held together by strong bonds, and strongly bonded atoms, like tightly racked billiard balls, are sensitive to the motion of their neighbors. Given, then, that the speed of light is exceedingly great, one would expect the ether to be harder than a diamond. But how could this be, since at the same time the ether appeared to be yielding, offering no resistance to the atoms, birds, and planets that passed through it?

The ether proved as difficult to verify experimentally as it did to describe theoretically. If the movement of light through the ether is comparable to waves rippling across the ocean's surface, then the perceived speed of light should be a function of the motion of the observer. Thus, a physicist in a boat traveling with an ocean wave would clock the wave's motion at a lower velocity than a colleague moving in the opposite direction. In the same way, how fast light appears to be traveling through the ether should depend on whether the observer is moving with the light or against it. Experiments were devised to measure this difference, but they all came to the same conclusion: the effect of the ether was undetectable; light traveled at the same speed no matter how it was measured. Finally, in 1905, Albert Einstein declared simply that the ether does not exist, and that light waves, in contrast with other oscillatory phenomena, propagate without an underlying medium. The vacuum was empty again.

But not for long. In 1928 the British physicist Paul Dirac combined the newly invented quantum theory, which applies to atomic phenomena, with Einstein's theory of relativity, which describes very fast motion, to give a complete description of electrons. At a certain juncture in his calculations, he had to take the square root of a term that described the energy of the electron. For the same reason that the square root of four is either $+2$ or -2, Dirac was confronted with a pair of solutions to his equation—one corresponding to positive energy, the other to negative. (Of the two basic kinds of energy in the universe, kinetic energy, that of movement, is always positive, whereas potential energy, that of position—say, where an object is located relative to an electromagnetic field—is frequently negative. In naturally occurring atoms, the sum of these two, called the total energy, turns out to be positive.) Dirac's positive-energy solution applied to electrons, whose motion he had set out to describe, but to what did negative energy refer? Some of Dirac's colleagues recommended that he ignore the seemingly absurd solution, but to do so would have been neither mathematically nor physically justifiable.

Rather than dismiss the strange solution, Dirac came to two conclusions. First, he predicted that there must exist in nature another particle, similar to the electron in mass but carrying a positive instead of negative charge (not to be confused with positive and negative energy). When, a few years later, the positron was discovered by the American physicist Carl Anderson, Dirac's faith in his formulas was vindicated. But his second conclusion was more troublesome.

Electrons circling the nucleus of an atom invariably seek the lowest energy level available to them, just as water always flows downhill. If, for example, an atom loses an electron, the vacant spot is filled by an electron from the energy level just above it. But if Dirac's theory about negative energy was correct, an infinite number of energy levels would exist below zero, and all electrons would eventually rain down into these levels. Given what we know about the behavior of electrons in atoms, the idea is incomprehensible.

In response to this dilemma, Dirac declared that the reason electrons don't fall to negative-energy levels is that those levels are already occupied; all of space brims with electrons that have negative energy and are incapable of interacting with ordinary matter. Rather than being surrounded by nothingness, everything in the universe is bathed in an imperceptible but ubiquitous ocean of fallen electrons known as the Dirac sea. There is no vacuum.

Not everyone was convinced, however. Apparently ill at ease with an assumed infinity of undetectable particles, the American physicist Richard Feynman, for one, found the Dirac sea unacceptable. But his solution to the problem was, characteristically, even more radical. Feynman recognized that the term describing the electron's energy in Dirac's equations was often multiplied by the symbol for time; instead of E, it was Et. Suppose, then, he suggested, it isn't energy but time that is negative. After all, the product Et doesn't distinguish between these two interpretations, because $(-E)$ times $(+t)$ is the same as $(+E)$ times $(-t)$. In this view, Dirac's negative-energy solution would correspond to ordinary electrons' running backward in time.

Strange as this suggestion might seem, Feynman took it one step further. Imagine a movie that depicts an electron near an atomic nucleus. Because it carries a negative charge, the electron is attracted to the positively charged nucleus. If you run this movie backward, the electron races away from the nucleus as if it had been repelled. And that's just how a positron behaves. Feynman concluded that Dirac's negative-energy solutions describe electrons traveling backward in time, which we experience as forward-moving positrons. Negative energy levels do not exist in this model, nor do the fallen electrons that Dirac used to fill those levels.

Draining the Dirac sea would have restored the vacuum to a state of emptiness, but for one new wrinkle. Feynman's interpretation works only in the context of quantum mechanics, whose rules—in particular, the uncertainty principle—dictate that certain variables such as energy cannot be pinned down precisely but fluctuate slightly. In the Feynman vacuum, a bit of energy can appear at random in the form of an electron accompanied by a positron (a pairing that preserves the electrical neutrality of the vacuum). An electron-positron pair materializes out of nothing, only to disappear without a trace. Here is a nothingness, then, that cannot be said to be absolutely empty, a nothingness in which "virtual" electron-positron pairs twinkle on and off like fireflies on a summer night. This new subatomic void, called the quantum electrodynamic vacuum, is a busier and friendlier place than the glacial ether or the fathomless Dirac sea.

It is also a place whose most peculiar feature has been confirmed experimentally. Imagine the void that lies between an atomic nucleus and the region in which electrons are found. If a virtual pair were to appear there, however briefly, the electron would be pulled slightly toward the nucleus and the positron pushed away. During its fleeting existence, the pair would, in effect, slightly alter the electrical character of the nucleus, reducing its charge relative to what it would be if the vacuum were truly empty. Not only has this reduction been measured, but its magnitude has been shown to agree exactly with the calculations—to an accuracy of one part in one hundred billion.

The construction of powerful particle accelerators has raised the possibility of other ways to probe the subatomic void. If the electrical force exerted by a nucleus were great enough, a virtual pair that bubbled up in its vicinity could be torn apart, with its electron pulled toward the nucleus and its positron propelled from the atom so quickly that the two would never have a chance to recombine and sink back into obscurity. An electron and a positron would then appear, as if from nowhere, and acquire a real, permanent, existence. Ordinary atomic nuclei do not carry enough of an electrical charge to test this prediction, but when two heavy nuclei, such as those of uranium atoms, collide at high speeds, they stick together briefly to form superheavy nuclei of sufficient charge to rend a virtual pair.

Searches for the spontaneous creation of real pairs out of the vacuum have been conducted for several years at an accelerator in Darmstadt, Germany. The experiments are difficult because the cloud of ordinary electrons and positrons that normally accompanies a violent collision of nuclei masks the effect. However, in late 1985, the group of German scientists at Darmstadt seemed to have succeeded: electron-positron pairs were observed in the vacuum. But beyond that, the pairs did not behave as predicted. The speed of the particles was expected to vary depending on the charge of the superheavy nucleus that pulled them out of the vacuum: the greater the charge, the more vigorously they should have been expelled. Instead, the pairs emerged from all collisions, irrespective of nuclear charge, at roughly the same velocity. Many solutions to the problem have been proposed recently, but one in particular deserves a closer look: something seems wrong with our idea of the vacuum.

In July 1986 Keh-Fei Liu, of the University of Kentucky, and Louis S. Celenza, Vinod K. Mishra, and Carl M. Shakin, of Brooklyn College, proposed a new vacuum state to explain the baffling invariance in particle speed. Specifically, they suggested that within the strong electric field that surrounds a superheavy nucleus the vacuum undergoes a phase change— the process by which the molecular configuration of a material alters abruptly and without apparent reason. Water, for example, when cooled to 32 degrees Fahrenheit, suddenly, and without the intervention of an external agent, turns to ice.

To imagine how this process might occur inside an atom, think of the quantum electrodynamic vacuum as a still lake from which drops of water, representing virtual pairs, bubble up in unpredictable places and at unpredictable times. To this picture add the possibility that the surface of the lake might occasionally be transformed into a sheet of ice. Particles might still pop up unexpectedly, but such slivers of ice are clearly different from water droplets. The equivalents of these particles in the new vacuum phase are called quasielectrons and quasipositrons. Theoretically, the behavior of a virtual pair of these objects would account for the unexpected results of the electron-positron experiments in Darmstadt. The more global question—how the vacuum changed phase, if, indeed, it did—remains unanswered.

With this new round of speculation, the vacuum has grown so crowded and cumbersome that we may again have to follow Parmenides' lead and reject the concept altogether, or, in the spirit of Einstein and Feynman, empty it of baggage once more. We needn't be deluded about the outcome of this second course. No doubt, were a housecleaning to take place, the vacuum would not remain empty for long. "Nature," said the Dutch philosopher Baruch Spinoza, "abhors a vacuum."

But perhaps he had it wrong. Nature, after all, is indifferent. It is *we* who abhor a vacuum, who recoil from the stillness of the void as from an open grave. Some deeply rooted urge compels us to fill it up, at least in our imaginations. We stuff it with ether, negative-energy electrons, virtual pairs, or quasiparticles—just as our ancestors filled it with a resounding, reassuring *om*.

5.

A Ripple in Gravity's Lens

The Velveteen Rabbit, by Margery Williams, is a story of how a boy's stuffed animal becomes real—not alive, but real. The story begins with the rabbit asking what that might entail: "Does it mean having things that buzz inside you?" No, replies an older, wiser toy, becoming real doesn't depend on how you're made. "It's a thing that happens to you. . . . It doesn't happen all at once. You become. It takes a long time."

Thus, in this classic tale, to become real is to become manifest—and thereby meaningful—to a human being. In due course the rabbit undergoes the process. Once a nondescript stuffed animal, indistinguishable from countless other stuffed animals, it comes to occupy the center of the boy's affections and is thereby transformed into something more than an artifact. The metamorphosis takes place not in the rabbit but in the boy's perception of it.

A similar shift is taking place among physicists in the way they regard one of nature's most fundamental yet elusive features: the gravitational field. Newton gave the first useful, scientific definition of gravity when he described it as action at a distance: all material objects attract each other, regardless of the medium, or lack of medium, between them. Newton devised a simple equation, the law of universal gravitation, which describes the attraction as the product of two terms. The first term depends on the mass of the attracted object, say an apple that falls to the ground. The second term is independent of the first, and it applies whether the falling object is an apple or a house; it is called the gravitational field of the earth.

Although it is invoked to explain such conspicuous phenomena as apples falling from trees, planets orbiting the sun and, indeed, orderly motion throughout the universe, the gravitational field of a massive object is itself a mathematical artifact. It has no more physical significance than the route map of an airline. Granted, just as the map enables one to measure distance or estimate travel time, the gravitational field helps the physicist predict natural events, and the predictions are extraordinarily accurate. But its predictive power does not make the gravitational field any less abstract.

Since Newton's day much has changed about how the gravitational field is understood. The chief architect of the change was Einstein, who in his general theory of relativity identified the field with space-time. Objects, he emphasized, are not drawn together by a force acting at a distance; instead they are simply moving along the curved lines of space-time, which is in turn warped by the presence of masses embedded in it. Although Einstein's radically different view of gravity endowed it with more power as a physical concept than it had in the Newtonian scheme, general relativity did not make the field more manifest. After all, space and time too are abstractions; they derive their meaning from the yardsticks and clocks that measure them. Whatever space-time may be "in itself," it is inaccessible to the senses, and so is the gravitational field to which it gives rise. Nothing that physicists have done since general relativity was introduced more than seventy years ago has altered that assessment.

Now, however, the gravitational field is becoming palpable—as real as ordinary fixed matter. Like the velveteen rabbit, the field itself is not changing; it is our perception and understanding of it that is being transformed. Driving this metamorphosis is the astronomical study of a phenomenon called gravitational lensing—the large-scale bending of light rays. In these observations, the gravitational field has finally become manifest as a shifting, rippling continuum extending throughout the universe, a field that bends, focuses, and otherwise distorts all the light passing through it. Such a medium must surely be called real.

The first detailed discussion of the bending of light rays by gravity appeared in the German journal *Astronomisches Jahrbuch* (Yearbook of Astronomy) in 1804. There the Bavarian geodesist Johann Georg von Soldner speculated that the path of a ray of light from a distant source would bend as it grazed a star on its way to earth. Von Soldner accepted Newton's view of light, that it is made up of innumerable minute particles, and found further support for his view in the writings of the Roman poet Lucretius, who in *On the Nature of Things* kept alive the Greek atomic doctrine of matter: "There is nothing which you can call distinct from body and separate from void to be discovered as a kind of third nature." The implication in this instance is that the light reaching the earth must be made of *something* (body) and therefore must be subject to the influence of the gravity of the star just as surely as a jet of water or a volley of cannonballs. It isn't necessary to know the composition of the particles to calculate their deflection because all objects, regardless of weight, fall at the same rate. For starlight grazing the rim of the sun, von Soldner applied the Newtonian concept of gravitational field to compute an angular deviation from a straight line of 0.84 seconds of arc, a value too small to be detected even by the best telescopes of the day. In any case, the sun is much too bright for stars to be observed next to its rim. Unfortunately for astronomy, von Soldner's paper was forgotten.

More than a century later, in 1911, Einstein, unaware of the previous calculation, published exactly the same result derived in the same way. But he added the ingenious suggestion that the effect could be observed during a solar eclipse, when the sun's brilliance is blocked and the stars near its rim become visible. The members of a German expedition to Russia to observe the total eclipse of August 21, 1914, were arrested as prisoners of war before they could make their observations, so the von Soldner–Einstein prediction could not be tested. It was just as well, at least in terms of the aims of science, for the predicted value was wrong.

In 1916, before the occurrence of another suitable eclipse, Einstein published a new calculation, derived from the general theory of relativity. The theory was based on a new concept of gravitational field—just the kind of concept Lucretius had rejected: neither vacuum nor matter, the field represents a "kind of third nature." In the revised calculation the angle of deflection of starlight by the sun was exactly twice as large as the earlier value. Crudely speaking, the extra deflection resulted from the influence of gravity on time, which had not been considered before.

Although it seemed slight, the difference was significant, not only because it distinguished Einstein's theory of gravity from Newton's but also because it made the deflection easier to measure. Three years later, in 1919, observations of a solar eclipse in Brazil and on the island of Principe, off the western coast of Africa, proved that Einstein was right about the effects of the gravitational field. Several dozen stars, whose true positions were known, appeared displaced in the sky by the predicted amount when their light passed near the sun. Overnight Einstein became a folk hero, and space-time a household word.

Armed with nothing more than Einstein's simple formula, one can in principle determine the influence of any star or galaxy on the light beams that crisscross the universe. In practice however, the distortions are so minute, and the accompanying calculations so forbidding, that the effect until recently had been observed only near the sun. Such limitations are now falling away. With modern telescopes, equipped with sensitive electronic and photographic detectors, and powerful computers, which dramatically speed up the analysis of astronomical data, scientists are able to construct a clearer picture of large-scale gravitational phenomena. The sky, it seems, is much livelier than it has ever seemed before. The gravitational fields of objects between the earth and distant sources of light do more than bend the light; they act as vast irregular lenses that distort the rays in a number of ways. Sometimes light simultaneously wraps around each side of the intervening mass, projecting two images of the same object. This effect was first detected in 1979, when a double quasar was discovered. The light source was actually a single quasar, a distant object thought to derive its immense power from a black hole at its center; the lens was an intervening galaxy.

Even stranger than the double image is the image of a source that happens to lie precisely behind the compact massive object. In that case the image of the source is a ring that surrounds the compact object. Although it seems extraordinary, the effect can be reproduced with commonplace objects: On a sheet of white paper, draw a heavy black dot to represent the distant source of light. Place an empty wineglass on the paper so that its base is centered on the dot. If you look through the top of the glass or peer around the bottom of its stem (and if you move the glass around a bit), you will discover a circular image of the dot.

Einstein himself described the cosmic version of this effect in a note to the journal *Science* in 1936, concluding with the remark that "there is no great chance of observing this phenomenon." But half a century later, the French astronomer Geneviève Soucail and her co-workers of the Observatoire de Toulouse and, independently, Vahé Petrosian of Stanford University and Roger Lynds of the Kitt Peak National Observatory discovered vast luminous arcs that have since been interpreted as partial Einstein rings. Such rings, and the crescent shapes that appear when the source is not perfectly centered on the compact lensing object, are special cases of optical images called caustics, or extended, rather than pointlike, focal regions. A familiar example of a caustic is the bright cusped shape that appears inside a wedding band when it is placed on a flat surface and exposed to oblique light.

Predicting the caustics likely to arise though gravitational lensing is straightforward in principle. First one assumes that masses of all kinds are distributed throughout the foreground. Then one traces the path of a simulated ray of light from, say, a background quasar through the intervening masses. At numerous points along the path, one computes the deviation caused by lensing, which enables the next leg of the path to be determined. The process is repeated until there are enough rays to build up an image that can then be sought with a telescope.

Ray tracing is the method Descartes used in the seventeenth century to explain the nature of the rainbow. For the sake of argument, Descartes imagined a spherical raindrop struck by a beam of sunlight made up of a large number of parallel rays. Applying the laws that govern the reflection and refraction of light by water, he mapped the rays as they entered the drop, bounced off its back surface, and exited from the front. By the time they emerged, he noted, they had become differentiated into their constituent colors.

Because the light from quasars comes from all directions, and because the deflecting lumps of matter are assumed to be randomly distributed in the universe, Descartes's procedure would be impossible to carry out by hand. But supercomputers render it tractable. Caustics resulting from the gravitational effects of complex distributions of matter have been extensively simulated, and it turns out that they can have the shapes not only of rings, crescents, or cusps but also of ovals, diamonds, clover leaves, or garlands. And the evidence for such images in the sky is beginning to accumulate.

If one could see with the unaided eye what is becoming clear in modern telescopes, the night sky would not be the inverted black bowl sprinkled with pearls we are accustomed to. It would be much brighter, and would resemble the surface of a lake rippled by the wind and reflecting the lights of a distant city. It would be a jumble of dots and streaks and patches of light in a richly textured tapestry. Examining an image, one would be at a loss to know whether it owed its shape to the original source of light or to the details of the intervening gravitational field. The only difference between the appearance of the surface of the lake and that of the night sky is that the former depends on reflection and the latter on refraction.

The gravitational field, and therefore the patterns it traces in the sky, is not static. The earth, stars, and galaxies move with respect to one another. Additional variation may be introduced by gravitational waves, which ripple through space like rollers across the ocean. Gravitational waves, which are analogous to light waves, are a consequence of general relativity; if the theory is correct they must exist. To date there is only indirect evidence of their existence: the inward spiraling of neutron stars that revolve about one another can be attributed to their loss of energy by the emission of gravitational waves. Nevertheless, physicists and astronomers firmly believe that these waves will soon be detected in the laboratory. Since they represent oscillations in the strength of the gravitational field, the waves affect the field's lensing action.

The recognition of this link between lensing and gravitational waves has led to provocative speculation. Bruce Allen of the University of Wisconsin in Milwaukee recently suggested how gravitational lensing might be exploited to detect gravitational waves. He discussed Quasi-Stellar Object 0957 + 561, a quasar whose image is doubled by a galaxy. The brightnesses of the two images grow and fall in similar patterns, but the variations of one lag behind those of the other by 415 days. Most astronomers attribute the delay to a difference in the path lengths of the two rays, but Allen proposed an alternative explanation.

Suppose, he argued, that a gravitational wave happens to be passing across the mass that is causing the gravitational field and acting as the lens. Suppose further that the wave makes a crest near one of the rays of quasar light and a trough near the other ray. At that moment the lens resembles the base of a wineglass that is denser on one side than on the other. Because a strong gravitational field slows down a beam of light just as thick glass does, the arrival of two rays from the quasar will be shifted in time. Thus a measurement of the time delay can reveal the presence of a gravitational wave. Allen found that the method works even after all the conceivable configurations of crests and troughs have been taken into account. He titled his paper "Gravitational Lenses as Long-Base-Line Gravitational-Wave Detectors," as if the lenses in question were bits of polished glass, not distant objects of cosmic proportion.

But gravitational lensing transcends its usefulness for the detection of gravitational waves, for it implies that the gravitational field can, in an important sense, be apprehended. How does a diver perceive the water that envelops him? Looking up, he sees its distorting effect on the light that arrives from the sky, and thus, even if the water is perfectly clear, he can see it. Of course, the visual evidence is corroborated by the evidence of the sense of touch: he can feel the water pressing against his skin.

The gravitational field is entirely analogous: We can see it by the way it refracts the light that comes through it, and we can feel its force on our bodies. The field encloses and holds us like an ocean of water, and to that extent it has become amenable to comprehension; it has become real. What's more, the process is irreversible. "Once you are real," as the velveteen rabbit learned, "you can't become unreal again."

6.

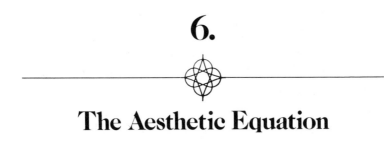

The Aesthetic Equation

Late in May 1925 the young German physicist Werner Heisenberg was staying on the island of Heligoland, in the North Sea, in an effort to rid himself of a severe case of hay fever. He was working on a radically new way to calculate the energy levels of atoms, taking into account not only the laws of classical mechanics but also the recent discovery that, in the atomic realm, energy comes in discrete bundles called quanta.

One evening he decided to check whether his values for the energy levels obeyed the law of conservation of energy—as they must in any viable theory. (The law holds that energy can be neither created nor destroyed.) Later he described what happened:

> When the first terms seemed to accord with the energy principle, I became rather excited, and I began to make countless arithmetical errors. As a result, it was almost three o'clock in the morning before the final result of my computations lay before me. The energy principle had held for all the terms, and I could no longer

doubt the mathematical consistency and coherence of the kind of
quantum mechanics to which my calculations pointed. . . . I had
the feeling that, through the surface of atomic phenomena, I was
looking at a strangely beautiful interior.

And then, too exhilarated to sleep, Heisenberg walked to the
beach, climbed atop a rock and waited for the sun to rise.

Contrary to the standard picture of how science progresses,
Heisenberg did not depend on experimental evidence to vali-
date his theory; that would come later. What convinced him
he was on the right track was the elegance, coherence, and
inner beauty of his approach—in other words, its aesthetic
qualities. Ten years earlier Albert Einstein had developed his
theory of gravity in a similar manner. In general relativity, as
the theory came to be called, gravity is described in terms of
the geometry of the four-dimensional space-time in which we
live. Einstein postulated his equations on aesthetic grounds:
they were simple and consistent and, like a work of art, they
felt right. "Anyone who fully comprehends this theory cannot
escape its magic," Einstein wrote when he announced his
creation in 1915. Experimental proof for the theory was almost
entirely lacking at the time. Only one piece of evidence, a tiny
anomaly in the orbit of Mercury, served as factual anchor for
Einstein's bold speculation.

Today elementary-particle physicists face an experimental
predicament even worse than the one Einstein did. Since the
middle of the century, theorists have been guided by the data
extracted from large-particle accelerators: the physical properties of the particles materialized in collisions of atomic nuclei,
the directions and speeds of the emerging fragments, the
changes in the processes as the energy increases, and much
more. When the particles were grouped according to their

masses, electric charges, and other attributes, orderly patterns appeared and pointed the way to other particles. The mathematical descriptions of these patterns led in turn to the prediction that more fundamental building blocks should exist. These too were found and named quarks. Theory and experiment inspired each other and kept pace. But eventually accelerator experiments became so time-consuming and expensive that theorists overtook their experimental colleagues.

Now theory is far ahead of experiment; indeed, it is out on a limb. The latest candidate for a unified description of all forces and matter, the so-called string theory, is so far removed from experimental testing that it has been called a theory of the twenty-first century accidentally discovered in the twentieth. The theory simplifies physics by assuming that all observed particles are different manifestations of the same fundamental entity. It also predicts, however, that this merging of identities cannot be observed except under conditions of heat and pressure that are inaccessible to current technology or, for that matter, to that of the foreseeable future. There is far less evidence for string theory today than there was for general relativity in 1915.

Proponents of the theory justify their creation by pointing to its elegance, coherence, and beauty. One of the attractions of the theory, for example, is that it combines general relativity with quantum mechanics. There is no law of nature requiring that these two theories, one for the microworld and the other for the universe at large, to fit into a single mold. But the twin hopes for unity and simplicity, both aesthetic criteria, are so strong that theorists have pursued a quantum theory of gravity for more than fifty years—so far in vain. No wonder they are impressed with string theory, its remoteness from the laboratory notwithstanding. But beauty, not utility, is their guiding principle, and critics who insist that physics should not traffic with untestable propositions dismiss string theory as recreational mathematics. In reply to this calumny, string theorist David Gross of Princeton University quotes the late Nobel laureate Paul Dirac, a founder of quantum theory: "The research worker in his efforts to express the fundamental laws of nature in mathematical form should strive mainly for mathematical beauty."

None of this sounds a death knell for experiment, of course. Wherever experimental evidence can be coaxed out of nature, it suffices to corroborate or refute a theory and serves as the sole arbiter of validity. But where evidence is sparse or absent—as it is for a growing number of questions in physics—other criteria, including aesthetic ones, come into play in an essential way, both for formulating a theory and for evaluating it. In view of this fact, it is imperative that physicists know what they mean when they make appeals to such standards as elegance, coherence, and inner beauty. Many professional scientists use these terms to refer to their work, but few take the trouble to define them. What, then, is meant by elegance? by coherence? And what is beauty, in the context of mathematical formulas and physical theories?

One man who has thought deeply about these questions is Subrahmanyan Chandrasekhar, a professor emeritus at the University of Chicago. His name is as well known within the physics community as it is unknown without. His main achievement is the proof that stars with more than 1.4 times the mass of our sun collapse when they die, whereas lighter ones escape that fate. (We now know that the collapse gives rise to a neutron star or a black hole.) This work dates to 1931, when Chandrasekhar was in his twenties and beginning graduate study at the University of Cambridge. But it was not only because of one spectacular feat of mathematical reasoning that he won the Nobel Prize. Chandrasekhar was honored for a spectrum of contributions to astrophysics, ranging from a complete description of stellar and planetary atmospheres to a mathematical theory of black holes.

In an autobiographical note to his Nobel lecture, Chandrasekhar described how his career has followed a highly ordered pattern. He would carefully pick a topic that suited his tastes and abilities, and spend several years studying it in depth. Then, when he felt he had achieved a perspective of his own on the entire subject, he would summarize it systematically in a "coherent account, with order, form, and structure." In this way he produced the seven monographs that constitute his scientific oeuvre.

Throughout his life Chandrasekhar has been preoccupied with the role of aesthetics in physics. In the introduction to his first book, a treatise on the theory of stellar structure that has served as the standard in its field for half a century, he struck the chord that was to reverberate through all the rest of his work. The book begins with an exposition of thermodynamics, the science of heat. This topic had been described countless times, and another summary seemed superfluous in a text on astrophysics. But Chandrasekhar had his own point of view. The treatment of thermodynamics he presented, a highly structured mathematical formulation, had been developed recently in Germany and was new to the English-speaking world. In Chandrasekhar's words this was, of all the treatments of the subject, "the only physically correct approach to the second law." And then he went on to remark that the logical rigor and beauty of the exposition was to exemplify the standard of perfection that should be demanded of any theory, including his own.

Chandrasekhar returned to the subject of beauty again and again, not only in his technical work but also in speeches and essays spanning forty years. In 1987 the latter were collected and published under the title *Truth and Beauty: Aesthetics and Motivations in Science.* The book is not a monolithic "coherent account," and much of it is so technical it is difficult even for physicists to read. Yet in the end it succeeds: through examples, such as the story of Heisenberg's revelation, through the accumulation of quotations from scientists and artists, and through accounts of his own experiences, Chandrasekhar conveys the perspective he has attained in a lifetime of creating theoretical physics and reflecting on its meaning.

In the pivotal essay of the collection, Chandrasekhar sets himself the difficult task of defining mathematical beauty. (He declines to take the easy way out, as Dirac, for one, did with his remark that mathematical beauty is as indefinable as artistic beauty, but is obvious when you encounter it.) As the title of Chandrasekhar's book suggests, he appreciates the power of Keats's insight that beauty is truth and truth beauty, but he wants to be more specific. Among the qualities of mathematical beauty he singles out, the most compelling is a sense of proportionality, of relatedness of parts to one another, of orderliness. According to Chandrasekhar, Heisenberg put it best: "Beauty is the proper conformity of the parts to one another and to the whole."

This definition can be applied to works of art as well as to works of mathematics. It suggests a feeling of balance, of parts that resonate rather than clash, of a total unity that bears an appropriate relation to its components. But concepts such as conformity, proportion, and order—like elegance, coherence, and even beauty itself—remain vague. What makes some things conform to one another, while other things don't? Why does one part fit naturally and beautifully into the context of the whole, while another one seems out of place?

One suggestion that renders the whole discussion more concrete can be found in the 1956 book *Science and Human Values,* by Jacob Bronowski, creator of the television series *The Ascent of Man.* According to Bronowski, the concept that underlies conformity, proportionality, orderliness, relatedness, and unity—in short, beauty—is the simple notion of likeness. "All science is the search for unity in hidden likenesses," he wrote, and, borrowing Samuel Taylor Coleridge's definition of beauty as unity in variety, "science is nothing else than the search to discover unity in the wild variety of nature. . . . Poetry, painting, the arts are the same search."

Although artists and scientists may seek the same end, they employ techniques peculiar to their disciplines. Poets, for example, apply rhythm, rhyme, assonance, and alliteration to seduce their readers into discovering likenesses, either hidden or overt, in the work itself. In prose writing, the success of the structure, which may make use of symmetry, foreshadowing, and echoes, similarly depends on the ability of the reader to notice correspondences. Painting and sculpture display correspondences in color, form, shape, and texture. In music, likenesses are found not only in the rhythmic patterns of similar sounds but also in the more subtle relations among harmonious tones.

Such stylistic correspondences acquire meaning, and beauty, to the extent that they reflect correspondences in the content of the work itself. Bronowski illustrates the point with a line from Shakespeare. When Romeo finds Juliet in the tomb, and thinks her dead, he laments, "Death that hath suckt the honey of thy breath." The rhyme of death with breath, and the sixfold repetition of the *th* sound, sometimes silent, sometimes buzzing, are the tools of the poet's craft. But the power of the line derives from its message, the comparisons of death to a bee, of Juliet to a flower—hidden likenesses between vastly disparate things.

In physics the most primitive tool for expressing likeness is the equality of two numbers. Further, one can construct pure numbers out of measured quantities by forming their ratios. The ratio of my weight in pounds to my daughter's weight, also in pounds, is the number 3.8, with no units. Likenesses between different things can then be expressed as the equality of ratios: apples can be compared with oranges. The equality of ratios is to physics what rhythm is to poetry, and balance to painting.

Many of the great discoveries in physics ultimately boil down to equalities of two ratios. When Archimedes discovered the law of the lever, for example, he found that a balance beam is in equilibrium when the ratio of the weights is equal to the inverse ratio of the lengths of the lever arms. The equation expresses a likeness between two seemingly unrelated quantities—weight and distance—and until the law was found the likeness was hidden. Two thousand years later Galileo showed that the ratio of the acceleration of a ball rolling down an incline to the acceleration of a ball in a free fall is equal to the ratio of the height of the incline to its length.

This likeness is subtler and more deeply hidden than the law discovered by Archimedes and, for that reason, it is in some sense superior. In a similar way, works of art gain in stature, and are considered to be more beautiful, as the appearances they unify are more widely varied. Thus there is greater merit in comparing death to a bee than to, say, sleep, and more poetry in the metaphor of honey for Juliet's breath than, say, wind. A scientific theory is beautiful to the extent that the phenomena it explains are unrelated—or at least seem so.

Newton's first great discovery about gravity was his demonstration that the ratio of the acceleration of a falling apple to that of the moon is equal to the inverse ratio of the squares of the distance of the two objects from the center of the earth. To appreciate the unexpectedness of this equality, think for a moment how differently the various quantities are determined. The acceleration of an apple is measured in the laboratory, with rulers and clocks. The acceleration of the moon, on the other hand, is a highly abstract concept that involves watching that distant orb glide through the night sky, timing its return from day to day, determining its distance by some trick of celestial triangulation, deriving its orbital velocity, and finally computing the change in that velocity per second. The radius of the earth is measured either by astronomical means or by painstaking land-based geodesy. None of this has any obvious connection to falling apples. That these disparate numbers, when expressed in the appropriate ratios, should result in such a simple equation is a miracle that must have astonished even Newton. To this day the equation remains a beautiful result of theoretical physics and amounts to the first example of the unification of seemingly disparate natural forces—celestial and terrestrial gravity. In a sense, string theory is simply a continuation of the program of unification that began with Newton.

An example of likeness established in this century was Einstein's theory that the ratio of the energies of two particles of light is equal to the ratio of their frequencies. This connection is the crucial idea underlying the quantization of energy—the notion that the energy of a wave comes in discrete quanta. Einstein's postulated relation between energy and frequency was completely unwarranted from the viewpoint of classical nineteenth-century physics: the energy carried by an ocean wave, for example, depends on other factors besides frequency, such as its height; the ratio of the energies of two ocean waves is therefore not simply the ratio of their frequencies.

The frequency-energy relation proved to be a cornerstone of quantum mechanics. It was the first hint that waves (characterized by frequency) can be thought of as particles (characterized by energy), and vice versa. Einstein received the Nobel Prize primarily for this discovery, not his more famous theories of special and general relativity. The relation was experimentally verified by the American physicist Robert Millikan, and helped lead him, too, to a Nobel Prize. That simple equality of ratios turned out to be one of the most powerful, and beautiful, unifying concepts of modern physics.

Just as the techniques of poetry go beyond rhyming, the expression of likenesses in physics can be more elaborate than the mere equality of ratios. Instead of likenesses between two phenomena, there can be family relations among the collections of facts. A prime example is the periodic table, which expresses similarities and regularities between more than a hundred elements. When the elements are merely indexed by atomic weight, a list results, but when that list is arranged so that similar elements appear in vertical columns, like the days of the week on a calendar, the list becomes the periodic table. Thus shiny metals are listed below one another, like all Tuesdays, and so are the inert gases. The power of the table comes from its ability to display the likenesses: if there is no known candidate for some entry, so that a gap appears in the table, the atomic weight of the missing element can be inferred from its position in a horizontal row, and its physical properties from those of its vertical neighbors. In this way the exact properties of yet undiscovered elements have been predicted theoretically, and subsequently confirmed in the laboratory. The classification of subatomic particles and the discovery of new ones have proceeded in the same fashion. The beauty of both the periodic table and the classification of elementary particles reside in their power to expose hidden likenesses.

The search for order need not be confined to objects such as atoms and elementary particles. At the highest level of theoretical physics, it applies to formulas and equations as well. The mathematical expressions themselves become objects of contemplation. Like a collector sorting seashells, the theoretical physicist plays with equations, writing them out in different forms and combinations, looking for likenesses until an orderly and pleasing arrangement is found.

In the nineteenth century, James Clerk Maxwell noticed that the four known equations of electricity and magnetism almost—but not quite—formed a striking pattern. When he exchanged the symbols for the electric and magnetic fields, he got back almost the same equations he had started with. To satisfy his aesthetic sensibilities, Maxwell boldly modified the last formula, making the symmetry perfect. The new equations showed him a profound hidden relation: an oscillating electric field gives rise to an oscillating magnetic field, and vice versa. Maxwell realized that this reciprocity could lead to a kind of bootstrapping effect, whereby the oscillating magnetic and electric fields would mutually sustain each other. It was a discovery that immediately led him to two important applications of the theory: his prediction of the existence of radio waves and his explanation of the nature of light. What started as an attempt to impose order and consistency on a set of equations ended with the most elegant and powerful theory of classical physics, and another giant step toward the unification of all forces.

In the final essay of his book, Chandrasekhar describes one of his own theoretical discoveries in order to illustrate the role of aesthetics in mathematical creation. The context is the general theory of relativity, and the subject is black holes. Einstein's equations describe exactly how space and time are intertwined and curved in the vicinity of one of these exotic objects. The black hole itself can be thought of as a star so compact that even light cannot escape the powerful pull of its gravity; to an outside observer the star is invisible. If the black hole is perfectly round and rotating about its own axis, the shape of space around it is described by just two variables: the distance from a point in space-time to the center of the star, and the latitude on the star to which the point corresponds. Time does not play a role, because nothing changes, and neither does longitude, because the black hole is symmetric about its axis of rotation. Under such conditions general relativity prescribes one equation in two variables.

Through years of study Chandrasekhar had become thoroughly familiar with the solutions of the equations describing the neighborhood of a black hole. He had also begun to look at colliding gravitational waves, about which much less was known. (Although astronomers generally agree that black holes exist, there is not yet any observational evidence for colliding gravitational waves.) Oddly enough, the interaction of two waves crashing head-on can also be described with one equation in two variables. If the waves are thought of as water waves approaching each other from opposite directions in a long, straight canal, only their positions along the canal, and time, play a role. Both waves can be conveniently measured from the time and location of their collision. Distance from the sides of the canal or above and below the waterline is not important. Thus to describe both round, stationary black holes and colliding gravitational waves, general relativity gives one equation in two variables. The difference is that in the former case both variables are related to position, whereas in the latter case one of them is time.

To his astonishment, Chandrasekhar found that if he transformed both equations through a series of abstract and formalistic tricks, they became not just similar but identical. The identity does not imply that the two physical phenomena are identical: how could a stationary black hole resemble two colliding waves? Instead the identity is only formal; the syntax of the two equations is the same, even though the interpretations of the two sets of symbols are entirely different. And yet, a hidden likeness had come to light.

A similar likeness appears in a much simpler context. If there is no air resistance, a cannonball shot horizontally from a cliff follows a path through space that mathematicians describe as half a parabola. The two variables of the problem are horizontal distance and vertical height, both measured from the cannonball's starting point. A ball rolling down an inclined plane is also described by two variables: the distance it has moved and the time elapsed. When the distance is plotted against the time, the figure that emerges is, mirabile dictu, also half a parabola. Just as in the case of general relativity, the two phenomena are dissimilar; the relevant variables include time in one case but not in the other, and the two identical mathematical descriptions have quite different physical interpretations. But the exposed likeness points to a beautiful and profound relation between the phenomena.

Chandrasekhar exploited the similarity he discovered between black holes and colliding gravitational waves to add a multitude of new and unexpected insights to the study of the latter—"implications," as he put it, that "one simply could not have foreseen." In his essay he emphasizes that only his sense of aesthetics, his firm belief in the beauty of general relativity, enabled this development. Without such a guide, and in the absence of experimental evidence, he would never even have imagined the transformations of the variables that made the two equations look the same. On the last five pages of *Truth and Beauty* are tables of mathematical symbols, quite impenetrable to the lay person, that describe black holes and colliding waves. But Chandrasekhar points out that the beauty of the scheme is apparent even to the untutored eye: "The pictorial pattern of this table is a visible manifestation of the structural unity of the subject."

It must be admitted that Chandrasekhar's monograph does not quite meet the standards of order, form, and structure that he set with the other, more technical chapters of his life's work. But then again, beauty, for all its importance in helping solve technical problems, is itself stubbornly resistant to codification. Perhaps the process is more important than the product, and the search for beauty more significant than its definition. If that is so, Chandrasekhar has succeeded brilliantly. By tirelessly seeking out hidden likenesses in some of the most abstract sectors of theoretical physics, Chandrasekhar has become the very embodiment of the quest for mathematical beauty in science.

7.

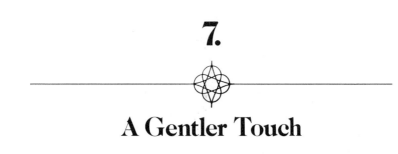

A Gentler Touch

In a book of alchemical emblems published in German in 1617, the epigram "Let nature be your guide" is illustrated with a picture of a serene young woman in flowing apparel who carries fruit and flowers through a moonlit landscape, leaving traces of her bare feet in a muddy path. Following a few yards behind her is a bearded chemist, equipped with a lantern and a pair of period eyeglasses that resemble swimmers' goggles. Nature, in this image, remains aloof, untouched by human hands. The scientist who would discover her secrets must study her from afar, relying for guidance on the clues she leaves as she disappears down the trail before him.

The passive role of the chemist implied by the print was vigorously rejected by the seventeenth-century statesman and philosopher Francis Bacon, who advocated active intervention through experimentation. Nature, Bacon believed, left to herself, is too inaccessible to reveal her mysteries. The deliberate manipulation of natural phenomena leads to understanding, because "in things artificial nature takes orders from man, and works under his authority." In mastery over nature, Bacon saw the key to knowledge—indeed, to human destiny.

Although Bacon was not a practicing scientist, his philosophy agreed so well with the experience of those of his contemporaries who founded modern science—the physicist William Gilbert, the anatomist William Harvey, and the astronomers Johannes Kepler and Galileo Galilei—that his ideas soon replaced the older, more contemplative attitude. For the next four centuries, scientists probed and pried into nature in an effort to uncover her innermost secrets. Anatomists dissected the human body down to the level of the cell. Chemists subdivided matter, both organic and inorganic, into molecules, and those, in turn, into atoms, until they could go no further. Physicists disassembled even the atoms, eventually smashing their nuclei into showers of elementary particles. In scientific research, reductionism carried the day.

The approach has been eminently successful. Today, we can manipulate cells and nuclei to achieve seemingly superhuman feats, from eradicating diseases to destroying cities. But even as insight into nature's workings is deepening, the limits of reductionism are becoming apparent. Some phenomena, it turns out, cannot be understood as mere sums of their parts, and others are altered or destroyed by the very act of analysis. Zoologists have discovered that gorillas in zoos and laboratories behave differently from those in the wild, so even the most meticulous studies of captive apes will not illuminate the true nature of gorilla life. Biochemists have found that the operation of organs within the human body is far too subtle and complex to be studied in vitro. And physicists have learned, to their astonishment, that they are unable to extend their quest into the realm of elementary particles; the quarks that constitute them refuse to be rent asunder.

Nature seems to have drawn an invisible line around certain phenomena—a line whose message is "Beyond this limit, destruction yields no new knowledge." Scientists, in response, are testing less intrusive methods of observation. Like the chemist in the old print, they are starting to drop back, to follow nature at a discreet distance; no longer intent on overtaking and subduing their subject, they are again letting her act as guide. In virtually all fields, including physics and engineering, scientists are once again learning how to explore the world without disturbing it. Ironically, this rejection of interventionism, which has served science well for four centuries, may turn out to be its biggest step forward.

Bacon's aggressive posture toward nature was in keeping with the temper of the Elizabethan age, in which hanging was a common form of punishment and torture an accepted method of criminal investigation. According to the nineteenth-century English essayist Thomas Macaulay, Bacon, who served as attorney general under King James I, was "among the last Englishmen who used the rack," implying that he adopted the same approach in his natural philosophy. Indeed, Bacon defined the aim of science as "putting nature to the question" at a time when one of the meanings of "questioning" was torturing. Bacon also characterized the mechanical arts, which we call technology, as clamping "bonds and handcuffs" on nature.

Those who distrusted the scientific enterprise argued that because Bacon's method requires a distortion of nature it is a poor way of gaining knowledge of the world, and this view persisted, despite the manifest success of chemistry, biology, physics, and all other disciplines, throughout the seventeenth and eighteenth centuries. Indeed, experimental science's most articulate and forceful critic, Johann Wolfgang von Goethe, was not born until a hundred and twenty-three years after Bacon was consigned to the English soil.

A man whose name is synonymous with the heroic attempt to gather all human endeavors under a single roof, Goethe was profoundly skeptical of any method for gaining knowledge that relied solely on taking things apart. Morphology, the science he championed, is the systematic study of forms and transformations as they are found unaltered in nature. The method he prescribed was contemplation; the aim, synthesis. Experiment and analysis, without which the empiricist would be lost, were of limited value to Goethe, because "truth, like divinity, is never to be known directly."

While there is much to admire in Goethe's desire to reconcile the contradictions of his time, we can be glad that his critique produced few converts among chemists and astronomers; otherwise, science would have ground to a halt (in fact, it did, in some quarters of Germany, where, during the nineteenth century, Goethe's speculative philosophy held sway). As it had since the days of Bacon, science progressed precisely to the extent that it intervened in nature through experimentation.

This is not to say that Goethe's outlook was ever banished from the intellectual arena. On the contrary, each generation produces at least a few skeptics who find in science something fearsome or deceptive. One of the most recent is the American poet Richard Moore, who, in the Spring 1983 issue of *The Georgia Review*, charged that science, to the extent that it relies on torture, is misleading, because such an approach tends to elicit the answers inquisitors desire rather than the truth. "Viewed in this light," Moore wrote, "science has been a game performed with the whole cosmos as the object of our own sadomasochistic delight."

Clearly, the moralistic thrust of Moore's argument is misdirected. When Bacon wrote about bonds and handcuffs, he was referring to the literal meaning of torture—of twisting or deforming—and only with respect to material, not people. According to him the principal method of science is forcing matter into artificial and stressful circumstances to see how it responds. This definition of the experimental method bears a close resemblance to the description of what musicians do with instruments (in Bacon's words, "The Bow tortureth the String continually, and thereby holdeth it in Continuall Trepidation") and, for that matter, what poets do with words. Moore is mistaken, too, when he says that putting nature to the question amounts to a self-fulfilling prophecy. The conceptual revolutions that have rocked science during the past four hundred years are proof of that.

But Moore is not completely off the mark in calling attention to the intimate relationship between the methods used by science to answer questions and the kinds of answers it elicits. Science, in its search for truth, is certainly intrusive, and frequently destructive. No one denies that such an aggressive approach affects the object of inquiry. The question is whether, by virtue of the inquiry itself, the object is obscured or transformed into something completely different. In recent years, a consensus has been building that that is exactly what takes place—that, at least in some instances, interrogation is an inadequate way of provoking responses from nature. So scientists have been searching for ways of teasing answers from her by gentler means.

One of the most impressive examples of noninterventionist science is computerized tomography, a technique, introduced in the early 1970s, that relies on a battery of detectors to collect data from a rotating X-ray beam as it passes repeatedly through an object. The composite scans—a series of cross sections, as it were, of the object's internal structure—are processed by a computer for three-dimensional display on a video monitor. As many as ninety thousand readings go into the development of a single image.

Because of its noninvasive nature, computerized tomography was first employed as a diagnostic tool in medicine. The technique provided physicians with a means for peering inside bodies without sawing bones and destroying tissue—and, most important, without damaging the very thing they wished to examine. As if by magic, they could gaze into a patient's head, lungs, and abdomen to search for tumors, hemorrhages, or signs of infection.

As anyone who has had a CAT scan can attest, the technique is the antithesis of torture. A brain-tumor diagnosis, for example, which used to require the skull's being sawed open, is now as simple and painless as taking a temperature. Since computerized tomography was introduced, other forms of noninvasive diagnosis, all based on the differing reactions of various materials within the body to electromagnetic radiation and other kinds of waves, have been developed, including magnetic resonance imaging, positron emission tomography, and ultrasound imaging. Equipped with the right lantern and spectacles, the modern physician can examine patients without so much as touching them.

Over the years, computerized tomography has also been used as an instrument of pure science, to probe a variety of solid objects that might otherwise remain inscrutable, including the earth itself. In tomography, modified to use seismic waves rather than electromagnetic radiation, geophysicists recognized a way of seeing what lies beneath the surface of the planet. Just as dense tumors change the passage of X rays through the body, various densities of the earth's crust, mantle, and core alter the speed and direction of seismic waves traveling from the planet's surface. Seismometers located throughout the world function as detectors, measuring the waves as they bounce back to the surface, then feed the information to a central computer, which creates a composite three-dimensional picture of the earth's interior.

Most of the noninvasive technologies that have come into use during the past twenty years grew out of discoveries in the physical sciences. Computerized tomography, for example, was developed by a physicist, Allan M. Cormack, of Tufts University, in Medford, Massachusetts, and an English electronics engineer, Godfrey N. Hounsfield (for which they won the 1979 Nobel Prize in physiology or medicine). Yet physics itself has been little affected by the new trend in instrumentation. Given that the science has been training its gaze on progressively smaller realms of matter, however, it was inevitable that it would also encounter a limit to intervention. Indeed, the limit had been anticipated: in the atomic realm, where quantum theory is sovereign, nature not only frowns on intrusiveness, it positively forbids it—an idea, known formally as the uncertainty principle, first proposed in the late 1920s.

The uncertainty principle decrees that certain properties of physical systems simply cannot be measured; in particular, whenever one pins down the position of an atomic particle, one thereby destroys all knowledge of its speed. A single particle's motion is thus much less certain than that of a normal, macroscopic object, which is characterized at all times by a definite position and speed. The uncertainty results from measurement itself.

In the ordinary world, position and speed are usually ascertained visually—that is, by means of light, which is much too feeble to influence a large object. In the case of a subatomic particle, however, the light transmits such a jolt that it knocks the particle out of its original position, giving it a new and unpredictable speed. In short, the measurement of one variable renders another unmeasurable. This limitation has been verified experimentally, but it did not pose any practical difficulties until recently. Now it is beginning to interfere with our understanding of the world.

Albert Einstein's theory of gravity predicts a phenomenon called gravitational radiation, which can be understood in analogy with light. When electric charges oscillate, they emit electromagnetic waves, which traverse air as well as empty space unseen and unfelt, until they meet other electric charges. When electromagnetic waves of a certain length encounter electrons in the retina of the human eye, the waves cause the electrons to oscillate, and the brain registers the resultant current as light. Gravitational waves are, according to the theory, similar, except that they interact with masses rather than with charges. In this view, the gravitational waves emitted by vibrating stars should cause other masses to oscillate. Physicists the world over have been searching for such oscillations in devices resembling seismometers, but they have been singularly unsuccessful.

The problem, it seems, is that gravitational waves are exceptionally weak. If a block of aluminum the size of a filing cabinet were exposed to such waves, it would move back and forth through a distance less than the diameter of an atom, and movement on such a small scale is governed by the rules of quantum mechanics: it is subject to the uncertainty principle. Thus, even if gravitational waves exist, and most physicists believe they do, they likely will escape detection, at least by any form of direct measurement.

In conventional gravitational-wave detectors, the speed of a wave is determined in the same way the speed of a fastball is found: by measuring the ball's position at two different times and dividing the distance traveled by the time elapsed. But the first steps in this sequence of operations are forbidden by the uncertainty principle, whereby measuring position alters speed, and vice versa. When the first detectors went into operation, about twenty years ago, their designers hoped that the effects of gravitational waves from outer space would be strong enough to escape the limitation set by the uncertainty principle. That hope has since vanished.

In response to this dilemma, Soviet, American, and Canadian physicists have invented a technique, known as quantum nondemolition measurement, in which the uncertainty principle is circumvented. To get an idea of how it works, compare the operation of an automobile speedometer: A rotating cable runs from the car's front axle to the dashboard. Attached to the axle is a magnet that exerts a drag on the axle's housing, which transmits this force to the cable. The strength of the drag (as represented by a needle) is proportional to the axle's speed of revolution and is therefore a measure of that speed. The important part of this scheme is that the position of the magnet need never be observed, let alone measured. Quantum nondemolition devices for detecting gravitational waves differ from automobile speedometers in that they monitor the frequency of oscillations rather than revolutions. Further, they are arranged so that the frequency can be determined without demolishing the detector's quantum-mechanical state (its precise microscopic configuration), allowing the measurement to be repeated many times. The next generation of gravitational-wave detectors will incorporate this technique.

While quantum nondemolition exemplifies the new approach to observation at the atomic level, an equally nonintrusive tack has been adopted, as well, at levels closer to ordinary experience—in engineering, for example. Cracks in airplane hulls, flaws in space-shuttle seals, corrosion in nuclear reactors: all are now being analyzed by means of X rays, laser light, infrared radiation, and ultrasound waves without the need to break or even dismantle the affected parts. Industry, naturally, is delighted with this technology, called nondestructive evaluation (NDE), because it makes inspection cheaper and more reliable, which leads to increased safety and efficiency, but just as significant are the scientific insights and the new applications that grow out of innovations in NDE.

One of the foremost laboratories for the study of NDE was developed by Joseph S. Heyman, at the NASA Langley Research Center, in Hampton, Virginia. The purpose of the laboratory, Heyman says, is to build a bridge between the needs of the real world of engineering and industry, on the one hand, and the fundamental insights of physicists and chemists, on the other. Most of the time these constituencies don't even speak the same language.

The difference between their ways of understanding matter became apparent not long ago, when the Association of American Railroads, in Chicago, asked the Langley laboratory to measure the internal stress in some samples of metal. Heyman, a physicist, had envisioned the samples as small chunks, about the size of sugar cubes. But that's not how the railroad business works. Sometime after Heyman agreed to the analysis, a truck pulled up to the laboratory and disgorged a couple of railway wheels, each four feet in diameter and weighing eight hundred pounds. A room had to be emptied and a special rack constructed to hold these monsters before they could be tested.

Heyman's nondestructive technique for evaluating internal stresses in metals makes use of two effects. A magnetic field applied to a sample reorients the molecules in different ways, depending on the direction and magnitude of the internal stress. The molecular configurations, in turn, affect the speed of ultrasound as it passes through the material. Measuring that speed provides a picture of the material's internal forces. Heyman found that one wheel had stresses that pointed inward, tending to keep the wheel together, and that the other harbored outward stresses that could, under adverse circumstances, cause it to break apart. The wheels were shipped back to Chicago for retesting, in an effort to check Heyman's novel technique. This time, the old-fashioned method of analysis—torture—was used: they were put to a saw. In the first wheel, the saw got stuck as the internal stresses closed the cut and clamped down on the blade. The second wheel exploded as soon as the saw nicked its rim.

Through the refinement of nondestructive techniques for evaluating materials on the microscopic level, scientists will be able not only to diagnose problems but also to prevent them. In the manufacture of composite materials, for example, such as epoxy laminates reinforced with graphite fibers, which are replacing aluminum in the construction of airplanes, it is crucial that the various layers of material adhere properly when they are baked. To this end, pressure and temperature in the oven are carefully monitored. But these properties provide only an indirect, and therefore less than reliable, measure of what really counts: the degree of bonding. By developing instruments that peer inside composite materials as they are being manufactured, and by understanding the details of how bonding progresses as a result of the application of temperature and pressure, safer airplanes, rockets, and cars will be built. Among the instruments that have already been developed as spin-offs of NDE research are a sensor for measuring the depth of burned human tissue, a gauge of bladder fullness for incontinent patients, and a device that monitors strain on mine bolts, which bind the posts that keep mine roofs from collapsing.

Where is all this headed? Noninvasive diagnosis, quantum nondemolition, nondestructive evaluation—the negative prefixes suggest conflict and the rejection of an established regime. That regime is not likely to be overthrown anytime soon. Much of what interests biologists, chemists, physicists, and geologists will continue to yield to invasion, demolition, and destruction, the martial arts of traditional science. But something new is afoot. We have entered a period during which man's relationship with nature—a relationship that in the West, at least, has been patterned after the intervention of traditional science—is being reevaluated.

The outcome of this new examination is not yet clear. Critics of science notwithstanding, a revival of passive contemplation in the manner promoted by Goethe is unlikely. By the same token, the old idea of mastering nature is beginning to seem unnecessary and even a bit foolish. As Bacon himself once said, "Nature, to be commanded, must be obeyed." That is surely the spirit that impels noninterventionist science, which has demonstrated that torture is not indispensable to empiricism. Perhaps in creating alternatives to the manipulation of natural phenomena, we not only chart a new path for science but also take a small step in the direction of wisdom.

8.

Einstein at the Ex

When I was a boy in Ottawa, many years ago, aside from the announcement that the Rideau Canal was safe for skating and ice hockey, the opening of the "Ex" was the most exciting event of the year. I suppose that the Central Canada Exhibition, as some people insisted on calling it, had serious commercial and agricultural purposes, but to us children it simply meant carnival, and carnival meant rides. I don't recall whether I rode the Wall of Death after taking physics in high school or in that same year, but in my mind the two are firmly linked. It was on the Wall of Death that I had my first conscious encounter with a fictitious force.

The wall was actually a cylinder ten to fifteen feet in diameter and about the same in height, with a floor but no roof. We stood inside, around the edge of the circular floor, waiting. As the cylinder began to spin we felt ourselves pressed against its side. Imperceptibly but relentlessly, the floor dropped away, leaving us glued halfway down the infernal pit, like flies on flypaper. It was wonderful.

I understood that what was holding us there was a combination of friction and something called the centrifugal force. What bothered me was that our physics teacher had insisted, with an enthusiasm bordering on zeal, that the centrifugal force is "fictitious." And there I was, hanging precariously on the wall, supported by a force that wasn't even supposed to be real!

By "fictitious" my teacher meant that the existence of the centrifugal force depends on the observer's point of view: the force can be made to appear or disappear by redefining the frame of reference. If my friend David, who, like me, later became a physicist, had watched from above as I rode the Wall of Death, he would have analyzed my motion without mentioning the centrifugal force.

David would have invoked the law of inertia: an object at rest tends to stay at rest, and an object set in motion tends to continue traveling in a straight line, unless acted upon by some external force. Thus, David would have said, at any given instant during my orbit within the Wall of Death, my body wanted to fly out of the cylinder on a tangent but was frustrated by the wall, which perpetually pushed it away from its desired direction and forced it around a circle. But from where I hung, there was no need to refer to the world outside, the world that stood still while the Wall of Death moved, the world into which my body wanted to be thrown. My world was this deep, dark cylinder, and something emanating from its center was, quite tangibly, pinning me against the wall.

The gut feeling that a good ride evokes is more memorable than any textbook description or lecture-room demonstration. Even the best course in physics misses one element in the understanding of mechanics: the feeling in our bodies of position and its changes—kinesthesia. Consider the two most significant variables for describing motion: velocity and its rate of change, acceleration. Of these, the first is seen and the second felt. If you are sitting in a big American car cruising on a smooth highway, or in an airplane flying high above the clouds, and close your eyes, you lose all sensation of speed. But as soon as the vehicle speeds up, slows down, or rounds a curve, you can feel the change; so, too, with carnival rides. Since the laws of mechanics are concerned largely with bodies that change speed, direction, or both, a visit to the carnival is ideal training for physicists.

The ride on which babies first experience the tickling of the stomach that accompanies changes in velocity is the swing. At the carnival, giant swings, with imposing names like Leonardo's Cradle, stop the heart during the moment of weightlessness at the end of each arc. The lovely to-and-fro of swings is related to the vibration of violin strings and the undulation of ocean waves, and is aptly called "simple harmonic motion." Contrasted with the rigidly uniform circular motion of the Wall of Death, simple harmonic motion appears effortless.

The most typical carnival ride is also the most commonly used example of carnival physics: the roller coaster. The regularity of the roller coaster is much more subtle than that of uniform circular or simple harmonic motion. Its motion is wild and tortuous, but when the motion is quantified, the conservation of energy emerges: the sum of the potential energy, which the car has by virtue of its distance above the ground, and the kinetic energy, which is proportional to the square of its speed, is always the same. Albert Einstein, in his only textbook on elementary physics, written with the relativist Leopold Infeld, chose the roller coaster as a perfect example of the conversion of potential energy to kinetic energy and back again. On the roller coaster, a budding physicist can *feel* the ebb and flow of invisible energies. What's more, she can chart the changes with her body; a graph of the potential energy is traced out by the ups and downs of the track itself.

Other attractions are less spectacular but no less revealing. Bumper cars obey the laws of collision—conservation of momentum and of kinetic energy—with bone-jarring fidelity. A steel ball, driven up a tower by the impact of a sledgehammer, steadily loses its momentum until finally a bell proclaims the prowess of the customer and the ultimate victory of gravity's inexorable pull. The hall of mirrors, in which curved reflectors distort the viewer's image into grotesque shapes, depends for its operation on the infinite repetition of the same fundamental principle of optics, the law of reflection. The simplicity of this rule, that the angle at which a ray leaves a mirror equals the angle at which it strikes, contrasts dramatically with the complexity of the images produced. The hall of mirrors thus exemplifies good science: explaining the complicated by means of the simple.

The loveliest ride of all, the Wave Swinger, combines elements of two perennial favorites, the merry-go-round and the swing. It consists of a gaily painted central tower with an umbrella roof from which seats are suspended by long chains. As the tower begins to revolve and then picks up speed, the seats swing out, moving in ever larger circles, their chains departing more and more from the perpendicular.

On the Wave Swinger you are subject to two competing forces: gravity, which pulls you down, and the centrifugal force, which pushes you outward. The law that governs this struggle, Newton's parallelogram of forces, is usually illustrated in high school laboratories by means of a little turntable equipped with strings, pulleys, and weights. At the carnival, you can duplicate this experiment and, indeed, enter into it; you are the weight, and the compromise between gravity and the centrifugal force is graphically expressed by the angle with which your chain departs from the vertical.

Careful observation reveals a remarkable phenomenon: the chains of all the riders depart from the vertical by the same angle. Heavier passengers do not hang more perpendicularly than light ones, and empty seats are never flung out beyond those that are occupied. As if by some conspiracy of nature the Wave Swinger's chains are always kept parallel.

The answer to this mystery is that both gravity and the inertia underlying the centrifugal force exert their influences in proportion to mass. A heavy body is indeed pulled toward the earth more strongly by gravity than an empty seat, but it is also pulled outward more by inertia. Einstein, when confronted with this fact, inquired further: Is it an accident, he wondered, that gravity and inertia depend on the same quantity?

Einstein's question ultimately led him to one of the most sophisticated creations of the human mind, his general theory of relativity. He began with a simple thought experiment: Consider free-fall—the experience of a parachutist hurtling through the atmosphere before pulling his rip cord. From the parachutist's point of view, there is no such thing as gravity. He doesn't *feel* the gravity that is so apparent when he stands on the earth, and he doesn't *see* its effects either. Indeed, he can take Newton's apple and place it in front of his nose, and it will remain there, immobile, relatively speaking. So long as the parachutist chooses not to expand his frame of reference to include the earth rapidly approaching from below or the air rushing past, everything he sees and feels can be explained without resorting to gravity.

Having demonstrated that gravity, like the centrifugal force, depends on the observer's frame of reference, Einstein went on to reveal a more fundamental connection between the two. He showed that gravity can be understood in terms of inertial forces, such as the centrifugal force—that inertia and gravity are really the same thing. This connection between two seemingly unconnected phenomena lies at the heart of the general theory of relativity.

Does this mean that the centrifugal force I felt while riding the Wall of Death was really due to gravity? If so, where did that gravity come from? Certainly not from the earth; centrifugal force also appears on the moon and on Mars and even in outer space. Around the turn of the century the Austrian physicist Ernst Mach proposed a solution to the problem. He believed that inertia comes from the collective gravitational pull of the countless distant galaxies that surround us. In his view, an object at rest tends to stay at rest because it is caught in a web of faint gravitational forces and centrifugal force is an attraction to the universe whirling in a circle.

Today Mach's conjecture is still an open question, hotly debated by cosmologists. I hope that it will survive in some form and be integrated into Einstein's theory because it is a deeply satisfying idea. If the distant galaxies, stars, and planets really do reach out and tug my body gently away from the Wave Swinger, then the universe is truly a unit. Its farthest reaches affect my body every time I accelerate in my car, turn a corner, or ride the Wall of Death. Pace my high school physics teacher, the centrifugal force is not only as real as gravity, it is an unbreakable link between me and everything else in creation.

Mass, motion, force, energy, relativity—the cardinal concepts of mechanics—were flamboyantly illustrated at the Ex, so it's a loss for physics that the traveling carnival is disappearing. Although some carnival rides have settled down in amusement parks, they are increasingly being replaced by larger, more frightening devices that lack the simplicity to illustrate physical principles. A modern roller coaster is not the primitive scaffolding of Einstein's time but a complicated machine whose loops and twists provide such a jolting ride that its mechanical principles are difficult to understand when observed from the ground, and impossible en route. Video arcades are also descendants of the carnival, but they do not teach science either; in a world where everything is possible, natural law holds no dominion. If that is where our children learn their physics, they will never acquire the ability to distinguish science from science fiction or the possible from the impossible. Worse still, they will be deprived of that visceral understanding of mechanics that we felt at the Ex.

9.

Indistinguishable Twins

On the fourth of June in 1924, Satyendranath Bose, a physics professor at the newly established University of Dacca, in East Bengal, wrote a letter to Albert Einstein, who lived in Berlin. "Respected Sir," Bose began, "I have ventured to send you the accompanying article for your perusal and opinion. I am anxious to know what you think of it." Bose went on to ask Einstein whether he thought the article worthy of publication, neglecting to mention that it had already been rejected by the prestigious English scientific journal *Philosophical Magazine.* But that fact scarcely mattered; Einstein replied immediately, informing Bose by postcard that he regarded the article as an important contribution. Indeed, Einstein was so impressed that he translated the paper into German himself and then arranged to have it published.

The article demonstrated that light can be understood in terms of photons, the particles of energy discovered earlier by Einstein, *only* if one makes a special assumption about their nature—that they are indistinguishable from one another. On the basis of this assumption, Bose derived a hitherto unsuspected property of photons: their habit of occupying the same state (in which their direction of motion, direction of spin, and energy are identical). Bose claimed that within the helter-skelter subatomic world, photons have a pronounced tendency to cooperate.

Because it is impossible to isolate and study individual photons, Bose's conjecture has never been verified directly. But predictions based on his claim have been borne out by experiment and, consequently, the cooperation of photons is now accepted as established fact. Other particles behave in the same way and, together with photons, are collectively called bosons, in honor of the man who identified the property they share. By cooperating in untold numbers, bosons transcend their submicroscopic insignificance and make their power felt in the everyday macroscopic world.

Two years after Bose sent his letter to Einstein, quantum mechanics, the modern mathematical theory of atomic processes, burst on the scene, creating a counterpoint to Bose's scheme—namely, that there is a second broad group of particles, which includes electrons, neutrons, and protons, whose behavior differs from that of bosons. Called fermions, after Enrico Fermi, the Italian-born physicist who first fully described their behavior, they never exist in the same state of motion but always compete.

This division of all subatomic particles into two broad groups—those that cooperate and those that compete—is more basic than the distinctions between the various kinds of particles themselves. And the idea from which the division is derived, the principle of indistinguishability, is more fundamental still. Understanding the role of indistinguishability in the structure of the world has led in recent years to the development of such powerful devices as lasers and superconducting magnets. But there is more to it than that. Since physics began, more than two millennia ago, the effort to comprehend matter has been dominated by a systematic search for its most basic constituents. The principle of indistinguishability has forced physicists to shift, or at least broaden, the focus of that effort.

The notion that subatomic particles are identical has surprisingly ancient roots. Its origin can be traced to the writings of the Greek philosophers Leucippus and Democritus, who created the atomic theory of matter during the fifth century B.C. in response to a nagging philosophical dilemma.

The aim of philosophy is to see the world as a unit, to understand it in terms of a single, all-encompassing principle. But the world appears not to live up to this ideal; our senses reveal an incomprehensible plurality of phenomena, a world that obviously consists of many things, not just one. By positing that everything consists of a vast number of identical particles too small to be seen or felt, the atomic hypothesis offered an ingenious compromise between the conflicting claims of theory and experience. Unity is to be found in the identity of the particles, and diversity in the complexity of their arrangements. That there in fact may be four types of atoms, as the Greeks believed, or even ninety-two, as we know today, does not alter the argument in any essential way. What counts is that there are only a few dozen different atoms, not trillions of them. Just as a handful of letters produces the infinite variety of world literature, a few dozen atoms, replicated countless times, constitute the world.

Modern physics not only retains the idea of the sameness of subatomic particles but actually sharpens it by introducing indistinguishability. When two ordinary objects, such as pennies or pins, are identical, they can still be distinguished by their positions relative to one another. We can refer to the one on the left and the one on the right, the upper and the lower one, or the one in front and the one in back. Thus, each object, even if it lacks intrinsic identifying features, is unique by virtue of its placement in space.

But in quantum physics, where the uncertainty principle prevails, exact position has no meaning. The paths of two electrons or two quarks, confined as they are within the tiny volume of an atom, are blurred to such a degree that it is impossible, even in principle, to tell the particles apart. They are more than identical, in the way that pennies and pins are identical; they are indistinguishable, and any formulation that refers specifically to electron number one and electron number two is wrong. The indistinguishability of subatomic particles has no counterpart in classical physics or in everyday life, but it rules the quantum world.

The key to understanding how indistinguishability leads to cooperation and competition lies in the collective behavior of bosons and fermions. Though contradictory, both behaviors refer to interactions between particles rather than to properties, such as mass and charge, that reside in the particles themselves. Interactions, in turn, are usually expressed as forces—such as the gravitational attraction between the earth and the moon or the electrical repulsion between two electrons—but the consequences of indistinguishability cannot be described in this way. Neither the tendency of bosons to act in unison nor that of fermions to compete is due to forces. The interactions are more fundamental even than that.

Nowadays, physicists presume that all the known forces in nature are mediated by particles—electromagnetism by photons, for instance, and gravitation by the hypothetical graviton. But cooperation and competition occur without mediation;

both behaviors are unaffected by the distance between particles, whereas the amount of repulsion between electrons or the
gravitational pull of one body on another is dependent on how
far apart they are. Beyond saying this, the only way to account
for indistinguishability and how it gives rise to cooperation and
competition is to examine the behavior of bosons and fermions.

Consider, as an analogy, the two ways that identical twins
respond to their intrinsic resemblance: some dress alike to
emphasize their similarity, while others dress for the sake of
differentiation. Indeed, the tendency of some twins to cooperate and others to compete is a universal theme. Castor and
Pollux, the original Gemini, lived in friendship until death
threatened to separate them because, of the two, only Pollux
was immortal. To reward the twins' exceptional fraternal devo-
tion, their father, Zeus, allowed them to spend alternate days
together in Hades, where mortals go, and on Mount Olympus,
the home of the gods. In effect, they shared one portion of
immortality between them, an act of extreme cooperation. The
biblical twins Jacob and Esau, on the other hand, fought even
before they were born. Crowded together in their mother's
womb, like quarks in a nucleus, they began a bitter struggle that
continued until they were grown men and had moved apart to
opposite ends of the country. The cooperative and competitive
propensities of separate sets of twins result from the interplay
of environmental and genetic influences, but how does the
dichotomy between bosons and fermions come about?

It is a matter of probabilities. With regard to bosons, imagine the four ways in which Castor and Pollux could be assigned
to their resting places: both in Hades, both on Olympus, Castor in Hades and Pollux on Olympus, or vice versa. In the first
two cases, or 50 percent of the time, they would be together,
doing the same thing. If, however, the brothers were not

merely identical but truly indistinguishable, there would be only three possibilities: both in Hades, both on Olympus, or one in each place. Such phrases as "Castor in Hades and Pollux on Olympus" would have no meaning. The twins would be together in two cases out of three, for a probability of cooperation of 66.7 percent. Applying this argument to two particles, indistinguishability increases the likelihood of their acting in unison by 16.7 percent. Though this may not seem to be a significant difference, in terms of probability, when trillions of particles are involved, it is enough to make laser light, for example.

The behavior of fermions can also be described statistically, but that is not how it was first approached. In the early 1920s, Wolfgang Pauli, then at the University of Hamburg, wondered why electrons, which are subject to the attractive force of the nucleus, do not descend to the atom's innermost energy orbit in accordance with their habit of seeking the lowest available energy state, just as water seeks the lowest possible elevation. In other words, what prevents an atom from collapsing like a ruined soufflé? Pauli ultimately concluded that electrons don't fall because they can't. The Pauli exclusion principle, as his theory came to be called, simply acknowledges that there is room for only one electron in each state.

Cast in statistical terms, the exclusion principle implies that the probability is zero that two electrons with identical spin will ever move simultaneously in the same direction at the same speed. And that is precisely what Fermi proved, using the wave equation, quantum theory's mathematical description of the wavelike nature of subatomic particles. He calculated the probability of finding any two objects together in one of two states and found three possibilities: If, like marbles and baseballs, the objects are distinguishable, the probability is 50 percent. But if the objects are subatomic particles and therefore indistinguishable, the probability is either 66.7 percent (bosons) or zero (fermions).

It may seem strange that indistinguishability leads to properties as contradictory as those exhibited by bosons and fermions, but analogous situations are easy to find. A good example is furnished by the two dominant forms of social contract in force in the world today. The same simple assumption—that all men and women are created equal—underlies both communism, which in its ideal form entails absolute economic cooperation, and egalitarian capitalism, which, in principle, establishes a system of unfettered competition in which everyone has an equal chance. The two systems couldn't be more different, yet they are based on the same assumption. And so it is with subatomic particles: the principle of indistinguishability yields a world divided into bosons and fermions. Though we have a long way to go before we fully comprehend why this should be so, we know that the world would be a radically different place if it were not.

If there were no fermions, all atoms would be alike—electrons circling nuclei in a narrow, dense band, similar to the rings around Saturn. But because electrons are fermions and must obey the exclusion principle, they fill progressively higher energy levels and thus travel in ever larger orbits as more of them gather about a nucleus. This explains why each atom in the periodic table is different in its configuration and, hence, in its physical properties. Simply put, the tendency of fermions to compete is responsible for the structure of everything from water molecules to human beings.

Fermions also help maintain the structure of the universe's large objects. While the sun's shape, and that of most other stars, is due largely to a delicate balance between the attractive force of gravitation and the outward pressure of the hot gases that stars are made of, there is a class of stellar objects whose shape depends solely on the competition of fermions: white dwarfs and neutron stars—tiny, cold corpses whose brilliance has faded and whose swirling gases have given up their struggle against gravitation. The only thing that keeps white dwarfs and neutron stars from collapsing is the exclusion principle. In the case of white dwarfs, gravitation has torn electrons from their atoms and compressed them to such a degree that any further compression would cause the electrons to violate their pact to avoid one another. Neutron stars are similarly compressed, except that for them gravitation is counterbalanced by the mutual avoidance of neutrons, which are also fermions.

Where the competition of fermions is responsible for much of the universe's diversity, bosons, through cooperation, provide power. Some of the most impressive demonstrations of their collective might are provided by man-made devices. Take the laser, for example. When a photon with a specific, pure color and definite direction strikes an atom, it may be absorbed, it may bounce off, or it may shake loose a second photon. It is the last of these possibilities, called stimulated emission, that the laser exploits. Because bosons cooperate, the second photon will be in the same state of motion as the first, and consequently possess the same color. Thus, where there was once one, there are now two, identical particles of light. If each of these, in turn, encounters another atom of sufficient energy, there will soon be four, then eight, and eventually a veritable avalanche. That's what is meant by "laser," which stands for light amplification by stimulated emission of radiation. The light that emerges from a laser results from the cooperative propensity of photons: its color is of an extraordinary purity, and since all of its photons travel in precisely the same direction, the beam spreads out much less than that of, say, an ordinary flashlight.

Utterly at odds in their effects, cooperation and competition would appear to be incompatible. But they can be combined. One of the most impressive examples is superconductivity— the disappearance, at low temperatures, of electrical resistance. Discovered in 1911 by the Dutch physicist Heike Kamerlingh Onnes, the effect was not fully explained until four decades later, and large-scale applications were rarely contemplated until superconductivity was found to occur in some materials at higher, more readily attainable temperatures.

The trick behind superconductivity is paired electrons. The partners in each pair, being fermions, cannot be in identical states of motion. In fact, they satisfy the exclusion principle by spinning in opposite directions. But as units the pairs act like bosons. Again, human behavior furnishes an analogy: Before people are married, they are, from an evolutionary standpoint, in sexual competition with everyone else; they behave like fermions. But as soon as they are wed, the competition ceases, at least in principle, and cooperation ensues; the couple behaves like a boson. In electrical conductors, coupled electrons are called Cooper pairs, after their discoverer, Leon N. Cooper, of Brown University.

Cooper pairs cannot form at room temperature, not only because electrons repel one another but because the jostling that occurs between the atoms of warm substances keeps them apart. Cooper realized that when the temperature is very low, however, and atoms become virtually still, electrons can exercise a hold on one another, albeit a weak one. This occurs when an electron pulls the positive nuclei of its surrounding atoms a little closer to itself than they would normally be. Thus, the electron surrounds itself with a cloud of positive charge, which, in turn, attracts a second electron, to form a couple.

Once a Cooper pair is set circling a wire loop by means of a magnet, all the other electron pairs, true to their bosonic nature, fall into step and follow the same motion. A current of Cooper pairs, unlike a current of unpaired electrons, will continue flowing unresisted forever. (All the pairs occupy the same quantum mechanical state of motion, and to disturb that state would require an influx of energy from an external source.)

There is a similar mechanical effect that has not yet found any useful application. Helium atoms, themselves constructed from fermions—two each of protons, neutrons, and electrons—tend to cooperate like bosons, as well. Consequently, at very low temperatures, liquid helium becomes superfluid. This means that a swirl in a cup of liquid helium will continue its motion forever because the unruly jostling of atoms, which quickly brings swirls in a coffee cup to a halt, is absent from the orderly march of bosons.

Though they may seem supernatural, superconductivity and superfluidity are natural consequences of indistinguishability; to realize that is to arrive at the true significance of this most basic of principles. By directing our attention to cooperation and competition, the most primitive interactions imaginable, indistinguishability has underscored the essential connectedness of the material world. It has compelled us to recognize that relationships between ultimate things are just as important as the things themselves.

10.

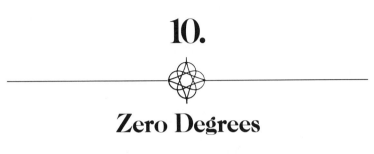

Zero Degrees

The other afternoon, in bright daylight, a crested cedar waxwing homed in on the reflection of a holly tree in a windowpane on my sun porch and crashed into the glass, breaking its neck. The accident underlined the insidiousness of invisible barriers: unexpected yet impenetrable, they are endowed with a sinister power until they are understood and brought under control. The waxwing's instincts, developed in the natural world, were rendered useless by the wall of glass, which lay well beyond the bird's range of experience.

Invisible barriers in the physical sciences don't break necks, but they do check human preconceptions in midflight and bring them crashing to the ground. The edge of the detectable universe—the cosmological horizon—is one such boundary, and the relativistic speed limit, beyond which no object can accelerate, is another. The consternation they cause stems from the fact that the extremely large and the extraordinarily fast lie outside the realm of human experience. A third such barrier is the absolute zero of temperature, the state beyond which matter cannot be cooled, in which all molecular motion ceases. Like an invisible wall, this concept jolts us out of a belief that we understand heat and temperature, and it demands deeper reflection.

Absolute zero as a theoretical concept was foreshadowed as early as 1699 in the work of the French physicist Guillaume Amontons, who conducted experiments in temperature and its measurement, but a century later it was still being dismissed by the majority of scientists. In 1804 the American adventurer-turned-scientist Benjamin Thompson told the Royal Society of London that "all attempts to determine the place of absolute cold, on the scale of a thermometer, must be nugatory." And his French colleague Joseph Louis Gay-Lussac called the notion "altogether chimerical." Not until the late 1800s was the idea revived, and today, when the finitude of the detectable universe and the speed limit of elementary particles are taken for granted, if not fully comprehended, the quest for the meaning of absolute zero, and the prospect of its experimental achievement, are at the forefront of experimental research.

The difficulty that the concept of absolute zero posed for nineteenth-century physicists stands in contrast to the ease with which its numerical value can be estimated from the observation that matter shrinks as it cools. In particular, when a quantity of air is kept at the same pressure and cooled from the temperature of boiling water (100 degrees Celsius) to that of freezing water (0 degrees Celsius), it loses about a quarter of its volume. If this trend were to continue, the volume would decrease to half by the time it reached −100 degrees Celsius, to a quarter by approximately −200 degrees Celsius, and to zero at about −300 degrees Celsius. Since a negative volume is inconceivable, the air's temperature cannot fall below that level. Although, in actual practice, air liquefies before it reaches absolute zero, applying the relationship between contracting gases and decreasing temperature—a valid mathematical exercise—yields the precise figure of -273.15 degrees Celsius, or -459.67 degrees Fahrenheit.

The first significant step in the direction of achieving absolute zero was taken in the 1870s, when a French mining engineer named Louis-Paul Cailetet liquefied oxygen at about -200 degrees Fahrenheit. Shortly afterward, he achieved the same result with nitrogen, which liquefies at -319 degrees Fahrenheit. Each of these advances provided low-temperature physicists with a new liquid coolant that could, in turn, be used to liquefy other gases. The challenge lay in devising bigger and better machines for maintaining sufficiently large and stable quantities of the critical liquids.

By the time Kamerlingh Onnes liquefied helium in 1908 at -457 degrees Fahrenheit, cryogenics laboratories had taken on the appearance of industrial plants, crammed with pumps and compressors, tubes and containers. Today, when temperatures of less than a degree above absolute zero are achievable, the trend continues at the University of Florida in Gainesville, where a new cryogenics laboratory opened in 1988.

From cylindrical pits in the laboratory's basement, three forty-foot refrigerators rise like guided missiles protruding from silos through the floors to the upper stories. Their surfaces are highly polished to reflect light and thereby insulate their contents. Separating their double walls is a vacuum, which, lacking a medium with which to conduct heat, provides additional thermal protection. To minimize motion (which is synonymous with heat), the cooling units are cradled by eighteen-foot pneumatic shock absorbers anchored in five-ton concrete blocks. The experimental area is surrounded by steel-and-cooper shields to screen out radio waves that generate minute electrical currents, which, in turn, create heat, and the rooms that house the refrigerators are soundproof. No one is allowed in the vicinity during experiments (the vibrations from a single footstep could raise the temperature significantly); all operations are handled by remote control.

The architects of this elaborate apparatus intend to cool a piece of copper to a millionth of a kelvin above absolute zero (a kelvin being the scientific unit of temperature and equal to nine fifths of a degree Fahrenheit). Then they hope to reach the holy grail of cryogenics—zero itself. If the Gainesville group does, in fact, fulfill the first of its aims, it will have exceeded the record established in 1987 at the University of Bayreuth, in Germany, by eleven microkelvins—a considerable achievement in the cryogenics world. But if they complete the next step, they will have accomplished the impossible, for the third law of thermodynamics, or the science of heat, expressly forbids an object to reach the state of absolute zero.

Thermodynamic principles are commonly stated in terms of restrictions, which are, for all intents and purposes, ironclad. The first law (also known as the law of conservation of energy) holds that energy can be neither created nor destroyed. The second law, which governs the direction of the flow of heat, dictates that during every transformation of energy a certain amount is dissipated, and therefore unavailable for doing work. Anyone who claims to have circumvented either of these laws (by inventing, in the first instance, a machine that creates more energy than it consumes or, in the second instance, a device that is 100 percent energy-efficient) would be considered a fool or a fraud. So it seems that in trying to circumvent the third article of the thermodynamic canon, the Gainesville physicists have embarked on a monumental misadventure.

But maybe not. The third law is not as well understood as

the first two; nor is it as firmly established. Indeed, in the light of modern developments in low-temperature physics, its dominion is more questionable than ever. And it is precisely the uncertainty that surrounds the third law—its potential for violation—that makes it provocative to scientists such as those at the University of Florida. The first and second laws, by virtue of their intellectual elegance and explanatory strength, dominate physics, but they are virtually barren as stimuli for experimental and theoretical research, whereas the prohibition against absolute zero raises fruitful questions about the ultimate nature of matter. Thus, the third law may turn out to be the Cinderella of thermodynamics, outshining her older, more illustrious stepsisters.

The first law, formulated in the 1840s by the German physicist Hermann von Helmholtz, is the easiest of the three to accept—in large part because it is the easiest to understand. When home appliances convert electricity into motion, heat, and light, energy neither increases nor decreases; it simply changes form. The same is true of conversions that, on first glance, seem unlikely—producing fire by rubbing together two sticks, for example (mechanical energy is transformed into heat, which, if vigorous enough, breaks atomic bonds in the wood, releasing chemical energy that assumes the form of fire). With every advance in our understanding of energy (including the discovery in the late nineteenth century of one of its most powerful forms, nuclear radiation, and, shortly thereafter, of its fundamental interchangeability with matter), the first law of thermodynamics gained strength. On no scale, from the subnuclear to the cosmic, has there been even a hint of a violation.

The second law of thermodynamics, at least in its modern form, is much more abstract than the first, but it is based on a commonplace observation: When two bodies—say, a pair of copper blocks—with unequal temperatures come into contact, heat always flows from the hot one to the cold one. In fact, the second law was conceived, in 1850 by the German physicist Rudolf J. Clausius, in just such operational terms, without a real understanding of the underlying mechanism. Nothing in the first law precludes thermal energy's traveling from a cold body to a hot one, making it hotter still; energy is conserved regardless of the direction in which it flows. The second law, therefore, stands on its own as a separate thermodynamic principle.

The meaning of this law becomes apparent when the movement of thermal energy is seen in terms of the atomic activity that makes it possible. In 1872 the Austrian physicist Ludwig Boltzmann identified atomic disorder as an essential thermodynamic quantity and restated the second law in its modern form: In any physical system, disorder increases naturally, and work is always required to reverse this trend. Imagine an orderly room in which toys are stacked neatly in a closet. A child enters and begins playing, scattering the toys with the greatest of ease. As parents well know, restoring the room's order will require greater effort than that expended to make the mess. The basis for this discrepancy is contained in the statistical nature of order: there are many more ways the toys can be spread around the room than ways they can be arranged in the closet. Considering the total number of possible configurations, disorder is simply more probable than order. The same is true of the molecules in, say, a cloud or even something as apparently stable as a block of copper. Indeed, the odds against molecular order in gases, liquids, and solids are so tremendously high that, as time passes, increased disorder is virtually certain.

Boltzmann's achievement was in showing how disorder, as expressed on the atomic level, relates to heat flow. He proved that the total amount of disorder in two bodies increases as heat flows from hot to cold. The essence of this demonstration lay in the realization that temperature is a measure of the speeds of molecules, that the molecules of a hot object jiggle about rapidly, carrying a lot of energy of motion, whereas those of a cold one move at a relatively sluggish pace. Think of a cold copper block placed below a hot one. At the outset, the jiggling molecules constitute two orderly piles—low speeds below, high speeds on top. Over the course of an hour, however, the positions of the blocks' molecules do not change appreciably but their speeds do. Within the blocks and particularly along the margin where they meet, warm, fast particles bump into cool, slow ones, transferring their energy. After several hours, high and low velocities are mixed together helter-skelter, top and bottom. Overall, disorder has increased.

The second law does not rule out the possibility of pushing heat uphill, as it were, from a cold object to a hot one, or of creating order out of disorder. It merely states that such a reversal of the natural flow requires an influx of energy—for instance, a steady supply of electricity. Proof is no further away than the kitchen, where a refrigerator draws heat from its interior and dumps it, via a metal grid affixed to the back of the machine, into the surrounding air.

Kitchen appliances cannot be used to demonstrate the third law of thermodynamics because it governs a much less accessible province—the realm of the very cold—than the first and second laws, which hold dominion over the entire span of temperatures. Doubtless, the exotic nature of this realm is the reason the third law was the last of the three to be proposed, in 1906, by Walther Nernst of the University of Berlin (for which he won the 1920 Nobel Prize in chemistry).

Physicists had previously assumed that, when cooled, molecules and atoms gradually slow down until they come to rest at absolute zero. But the quantum-mechanical theory of matter, which was emerging at the turn of the century, does not allow that: the lowest energy available to an atom, called its ground state energy, is not zero; there always remains a small, irreducible quivering that can be theoretically described and experimentally measured. The energy associated with this state cannot be shared with other objects (it cannot "flow") or be interpreted as heat, so it does not count as disorderliness. (Think of the child's room again, this time imagining that the toys consist solely of watches of various sizes and shapes. The turning of the watches' gears—their internal quivering—has no bearing on the room's order, on whether the toys are stacked in the closet or scattered across the floor.) Reasoning in this manner, Nernst arrived at a new understanding of absolute zero: rather than the absence of motion, it signifies the absence of disorder, or a state of perfect order—and this became the initial version of the third law of thermodynamics.

Six years later, while surveying the specific heat—the amount of heat lost when temperature drops by a single degree—of certain chemical elements, Nernst discovered a remarkable fact: During the approach toward absolute zero, or perfect order, each step is more difficult than the preceding one. With each successive removal of heat, temperature decreases less. Nernst speculated that this relationship is not merely an accidental property of particular substances but an attribute of all matter. Ascribing the phenomenon to the difficulty of corraling huge numbers of unruly molecules into a single state of perfect order (the chances are astronomical that at least a few will elude capture), Nernst proposed a stronger form of the third law: Absolute zero is unattainable.

Compare a similar injunction, in a different context: Einstein's claim that objects cannot attain the speed of light. When first proposed, this law seemed strange, because the phenomenon occurred in the absence of a specific, countervailing force. But later Einstein showed that, as an object accelerates, its mass increases (as summarized in the formula $E = mc^2$). From this he deduced that as an object approaches the speed of light, the energy required to further accelerate it approaches infinity. A change of variables, from speed to energy, made an incomprehensible limit appear comprehensible. In a roughly similar way, Nernst's switch from the absence of motion to the absence of disorder as the determinate of absolute zero made his decree seem plausible. The implication is that, as an object approaches a state of perfect order, the effort required to remove the remaining disorder approaches infinity.

About the reality of this barrier there is no doubt; it has been confirmed repeatedly in the laboratory. But whether an infinite amount of effort is necessary to reach absolute zero in all cases is less certain. Is the third law everywhere and always binding? That is the question lingering in the minds of the scientists in Gainesville.

That the question can even be raised goes back to the statistical nature of order. In principle, there is a small but finite probability that an object will reach a temperature of a billionth of a kelvin or even zero. For macroscopic systems, such as blocks of copper, consisting as they do of trillions of atoms, this probability is vanishingly small. A single atom, on the other hand, could readily jump into its ground state and, hence, into a state of absolute zero, if such a concept were applicable. But it is not. Temperature is a measure of the random motion—the thermal energy—of a group of particles, and is meaningless with a single atom.

Whatever doubts might be raised by the third law concern

matter that lies between the macroscopic and the microscopic scales, in particular, matter that approaches the small end of the spectrum, where the probability that random motion will cease is most favorable (fewer atoms means fewer ways that they can be disordered). In recent years, objects of this kind, called mesoscopic because of their intermediate status, have captured the attention of physicists. Containing a small number of atoms, and therefore visible only under powerful microscopes, they occupy a no-man's-land where electrical resistance, normally a fixed property of materials, shifts wildly; where a tiny disturbance in one side of the object can affect the other; where, in short, the ordinary laws of physics do not seem to hold. Chances are that mesoscopic objects are the correct size to bridge the gap between a microkelvin and zero (if the child in the nursery has only two toys to play with, she might put them on the right shelf in the closet by pure chance). It is on this slim statistical possibility that the Gainesville physicists pin their hopes.

Their refrigeration process, called magnetic cooling, is divided into two stages. During the first stage, copper samples are placed between the poles of a powerful electromagnet and chilled to the lowest temperature attainable by conventional means—by being bathed in liquid helium. In the crucial second stage the magnet is switched off, and within two or three days the temperature of the copper spontaneously falls to an even lower level.

Magnetic cooling is based on the principle that heat represents the random motion of an object's constituent particles. Since atomic nuclei behave like tiny magnets, they respond to an external magnetic field by lining up in parallel orientation—north to south and south to north. When the magnetic field is removed, the nuclei are jostled back into random orientations by collisions with their neighbors. Such collisions slow down atoms, so the temperature is reduced.

The process can be repeated again and again, but the technical difficulties mount with every step. A particularly insidious problem at temperatures approaching absolute zero is the minute amount of heat emitted by the residual radioactivity in construction materials. Fallout from the atmospheric nuclear tests of the 1950s, the 1986 Chernobyl accident, and other sources, now salts the entire surface of the earth, making it difficult to find metals that are free of radioactive debris or that do not become contaminated during the fabrication of machine parts when, in liquid form, they are exposed to the surfaces of their containers, as well as to the atmosphere. The Gainesville scientists hope this problem will be too small to affect their refrigerators.

Even under the best of conditions, the road to absolute zero will be neither short nor smooth. On the experimental side, the refrigerators will become larger, more complex, and more expensive, as the samples become smaller. The insulative measures taken will have to be heroic as the heat leaks grow increasingly difficult to perceive. On the theoretical front, the third law of thermodynamics must be reconciled at some point with the quantum-mechanical behavior of mesoscopic systems. The resultant formulation may lack the concision of Nernst's statement that absolute zero is unattainable, but it will touch on philosophical questions about the meaning of quantum mechanics (what is the relationship between randomness in thermodynamics and randomness in quantum theory?) and thermodynamics (will what we learn about the attainability of absolute zero have any impact on our view of the second law, which is also based on a statistical understanding of atomic behavior?) and about the process of measurement (what sort of thermometer will detect extraordinarily low temperatures without itself introducing heat?).

Work along these lines is just beginning. The third law could go down in the history books as a mistake. Or it could be dismissed as a trivial practical observation similar to the statement that no object can have infinite energy because there isn't enough available in the universe. A new phenomenon could be discovered in the submicrokelvin range that would render the whole discussion obsolete. Or, on the contrary, Nernst's proposition could, by virtue of its intrinsically quantum-mechanical nature, emerge as the most significant of all thermodynamic principles. Faced with so many possibilities, one is reminded of a comment made by Niels Bohr, the father of quantum mechanics: "The opposite of a correct statement is a false statement. But the opposite of a profound truth may well be another profound truth." Unlike the first and second laws of thermodynamics, the third law fascinates precisely to the extent that it might yet provoke an opposing truth.

11.

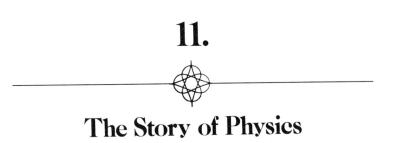

The Story of Physics

Tradition has it that the door to Plato's Academy, in Athens, bore the forbidding inscription, "Let no one enter who is not a mathematician." Geometry, for Plato, was the supreme example of pure reasoning—abstract, immutable, detached from human affairs, as close to absolute truth as one can get. Thus it provided the perfect intellectual training for philosophers. Accordingly, the academy's students—among them Aristotle, whose thinking was to dominate the physical sciences for almost two millennia—acquired a solid mathematical education.

But Plato wasn't the only teacher in Athens. Overshadowing his academy in prestige and popularity was the school of Isocrates, the orator. For more than fifty years, Isocrates taught the elite youth of Greece, and his influence on the great politicians, writers, and historians of the age was enormous. At the core of his philosophy stood *logos*, the word. Through language, he wrote, "we have come together and founded cities and made laws and invented arts." Language, in brief, is what made us human. Whereas Plato insisted on the study of mathematics, Isocrates emphasized rhetoric, literature, and history. Thus, even as science was beginning to stir in the womb of philosophy, it was faced with a choice of directions. Would it follow Plato along the path of mathematical idealism, or would it become part of Isocrates' literary and historical humanism?

At first the odds seemed to favor Isocrates. Physics, for instance, was largely narrative in the beginning. The Roman poet Lucretius cast the atomic theory of the early Greek philosopher Epicurus in verse form, in his celebrated poem, *On the Nature of Things.* In more than seven thousand lines of heroic hexameter, he attempted to deduce all the rich complexity we observe in nature from a handful of fundamental axioms. Though the poem did not anticipate the modern conception of atomic structure, it served to keep alive the doctrine of atomic materialism—that everything in the universe is composed of bits of primal matter—which is the basis for today's physical world-picture. And even much later, in the seventeenth century, Galileo published his fundamental mechanical and astronomical discoveries in the form of dialogues written in the colloquial Italian of his fellow citizens.

But the end of the literary mode was in sight; by the second half of the seventeenth century, the age of Newton, equations had replaced words as the vocabulary of physics. From that time until quite recently, physics concerned itself solely with quantifiable facts and the elucidation of their complex interrelationships, with no concern for how the universe came to be the way it is. The proper subjects of physics were considered timeless: one recounted the laws of motion by resorting to symbols and equations, not to such narrative devices as foreshadowing and denouement. Gravitation stood outside history; it would always be the same. Of course, the passage of time played an important role in such individual phenomena as the fall of an apple or the heating of a kettle, but not in the grand laws that governed such events.

The absence of history was particularly evident in the study of elementary particles, which began in the twentieth century. The goal of this research was to identify the ultimate building blocks of matter, measure their properties, and discover their modes of interaction. Everything anyone wanted to know about an atomic particle could be represented by a number. In the study of matter, it seemed, Plato had won out over Isocrates.

But now the situation has changed. During the past fifteen years or so, elementary-particle physics has merged with cosmology, the study of the structure and evolution of the universe. Official acknowledgment of the union came in 1982, when the *Review of Particle Properties,* the bible of high-energy physicists, included for the first time results derived from astrophysics and cosmology, a practice sure to grow in the coming years. This collaboration between the science of the unimaginably small and the science of the incomprehensibly large has made its way into popular literature, as well, via a small avalanche of books that started in 1977 with Steven Weinberg's *The First Three Minutes,* in which the Nobel laureate described the brief but formative opening act of the universe. Among the many outcomes of this union is one that is no less remarkable for being overlooked: the study of matter has become a natural history. Physicists have discovered that there is a story to matter, a story in the true sense of the word, with a riveting beginning, a middle rich in detail and bristling with surprising subplots, and a suspenseful ending.

James Clerk Maxwell, the nineteenth-century physicist who invented the theory of electromagnetism, once remarked, "It is when we take some interest in the great discoverers and their lives that [science] becomes endurable, and only when we begin to trace the development of ideas that it becomes fascinating." In this astonishing admission, one of the world's masters of abstract reasoning demonstrated that he, like the rest of us, was seduced by stories. What Maxwell had in mind could not be found in physics itself at that time but only in the history of physics. That's where the stories were.

The history of physics, of course, is not to be confused with the history of matter. The history of physics—or of any other science—is a discipline unto itself. Its practitioners are historians rather than scientists; its methods and aims are as different from those of physics as a library is from a linear accelerator. The history of matter, by contrast, implies a built-in historical component; the quality of temporality is to be found in the object of scientific inquiry rather than in the development of the discipline itself.

The best-known example of a science with a built-in history is evolutionary biology. At about the same time that Maxwell was composing books in the language of calculus, Charles Darwin was writing his revolutionary theses *On the Origin of Species* and *The Descent of Man* in ordinary English. They can be read as literature even though they are works of professional biology; literary devices—digressions, subplots, and embroideries—abound. Who can resist being fascinated by Darwin's entertaining tales of the engineering skills of bees, the love antics of bower birds, or the horrors of infanticide among the Polynesians?

But more important than the literary appeal of the genre was the introduction of history into biology. Before Darwin, biology was dominated by taxonomy, the static science of classification. Reinterpreted according to the theory of evolution, taxonomy acquired a dynamic, and classification was made to submit to the laws of lineage. By introducing the dimension of time, Darwin's theory connected a multitude of disparate taxonomic observations into a logical sequence of development, just as the trunk, branches, and twigs of a tree connect its leaves. A similarly significant shift occurred in geology with the advent of plate tectonics. Whereas the surface of the earth was once regarded as a stationary mosaic of landmasses and oceans, it is now understood to be a dynamic system in perpetual, turbulent change, a system whose history can be read in the rocks that exist today.

In recent years, physicists have found signs of impermanence in their own domain as well. Of course, no one has ever suggested that matter, in the superficial sense, is invariable. Food is digested, gas burns, trees grow; the study of such transformations is the business of chemists and biologists. But the substrate—the atoms that make up the pepper steak, Exxon unleaded, and the Douglas fir—was long considered immutable. And with the exception of certain rare and violent phenomena (radioactive decay is one), this view was correct, at least with regard to geologic time. To a chemist or a geologist, anything that keeps the same identity for six hundred million years—say, a normal atom of carbon—is for all intents and purposes immutable. But by dint of their concern about fundamental questions of origin, the cosmologist and the elementary-particle physicist are inclined to take a longer view; turning back the clock to nearly the beginning of the universe, they have found that the carbon atom led a quite different life before it entered its present period of stability.

The evidence from which this story has been woven consists of a vast collection of data from such sources as computer simulations, experiments with particle accelerators, and astronomical observations with radio telescopes, balloon-borne cosmic-ray detectors, and space probes. The carbon atom, for example, was constructed from three helium atoms in the interior of a star and then flung into outer space when the star exploded, long before the birth of our solar system. The helium, for its part, had been synthesized from hydrogen, either in the same star or in the big bang with which the universe began fifteen billion years ago, and the hydrogen, in turn, from quarks. Indeed, most of the drama in the story of matter is compressed into the opening moments of the universe, an episode so brief that it was overlooked altogether by earlier physicists.

Today we know what we would find if we could travel backward through that inaugural episode: When its age could be measured in minutes, the universe was bathed in light. Before that, it had been like the core of a hydrogen bomb—an inferno of cooking helium. The helium had condensed from a scalding soup of protons and neutrons that, in their turn, had been forged from quarks. Ever closer to the beginning of everything, we would find a mixture of equal parts of quarks and antiquarks, of matter and antimatter, and finally—when the universe was less than a nonillionth or 10^{-30} of a second old—a world in which the term *matter* ceases to have meaning because all the elementary particles we know (including quarks, antiquarks, electrons, positrons, muons, tau leptons, neutrinos, vector bosons, gluons, even particles of light) were continuously and rapidly changing into one another in a formless hurly-burly. As a dramatic story, this three-minute epic rivals the Book of Genesis and other creation myths.

But just as history is more than description, the story of matter is more than an entertaining yarn; an understanding of the past and present nature of matter, as well as the physical laws that govern relationships between the two, allows one to use history as a tool for both analysis and prediction.

A good illustration of analysis is the way that knowledge of matter's history contributes to an understanding of neutrinos. These ghostlike particles, produced in certain radioactive decays and in the collisions of nuclei in the sun and other stars, flit about the universe in immense quantities but don't appear to contribute to the constitution of matter. Until recently, one might have imagined the world little changed by their loss, just as one might imagine the world remaining more or less unaltered by the absence of mosquitoes. But the history of matter teaches us that the loss of these seemingly inconsequential particles would render the universe unrecognizable.

With the help of particle accelerators, three kinds of neutrinos have been identified. Sorting out the subtle differences among them is a complicated and time-consuming task that occupies hundreds of physicists and is far from complete. But cosmologists have determined that neutrinos were part of the primordial soup of particles that constituted the universe during the big bang. One of the clues to the state of affairs at the time is the amount of helium that can now be observed. Helium accounts for about a quarter of all the matter around us and, after hydrogen, is the most abundant element. The history of matter tells us that the amount of helium in the universe today depends critically on the number of types of neutrinos that have always existed: to the physicist, helium's share of the total chemical inventory, 25 percent, implies that the number of neutrino types is either three or four. If there were more types, the calculations reveal, the primordial stew would have been richer and hotter, more helium would have formed in the cosmic pressure cooker, and there would be more helium in evidence today. Conversely, fewer types of neutrinos would have yielded less helium than the 25 percent found today. Thus, cosmologists are able to say to elementary-particle physicists: Maybe you will find a fourth type of neutrino, maybe not. But don't bother to look for a fifth; we are certain it doesn't exist. Knowledge of the history of matter has placed a powerful restriction on the classification of elementary particles.

Whereas cosmological research into the universe's past elucidates questions posed by particle physicists, particle physics, in turn, helps cosmologists predict the future of the universe. Until recently, one of the cornerstones of the study of matter had been the assumption that protons—which, together with neutrons, constitute all the nuclei of all the atoms in the universe—never decay. Modern theories on the interaction of quarks, however, have raised suspicions about the proton's indestructibility. The issue is of no small consequence because, if protons do indeed decay, all forms of matter— mountains and oceans, the earth and the other planets, stars, and galaxies—will eventually crumble into a desolate gruel of neutrinos and electrons. (I'm leaving aside the unpleasant possibility that, before decay occurs, gravitation will stop the distant galaxies in their outward paths and pull them back together again, causing the universe to collapse upon itself in a cataclysmic reversal of the big bang.)

According to current evidence, if protons decay, they do so, on average, after an elapsed time that is longer than the present age of the universe. Whatever the exact lifetime turns out to be, it bears critically on the fate of matter. It is not surprising, then, that the suspected impermanence of the proton is now the focus of experiments the world over.

At the dawn of science, when philosophers began to speculate on the nature of physical reality, physics and cosmology were indistinguishable. Later generations of physicists, increasingly preoccupied with small-scale laboratory experiments of short duration, banished cosmology to a back room in the house of science, where it became a harmless, inconsequential pastime of astronomers. But the exile did not last. Just as an inquiry into the nature of man must concern itself with birth and death, so the study of matter has raised questions about *its* origin and destiny. Physics, in acquiring a historical dimension, has taken a step toward closing a long-standing rift. Equally important, the reconciliation is occurring in a way that is faithful to the teachings of both Plato and Isocrates.

12.

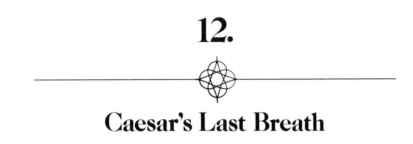

Caesar's Last Breath

With my morning coffee, I always study the weather map in *The Washington Post*. And when there's time, I glance at the jagged plot of the stock market index and at a few other data graphics—say, the performance of the yen against the dollar or the upward march of the defense budget since the Second World War. In the course of one year, I probably look at several thousand such pictures in newspapers and magazines, not to mention a few hundred more in physics journals. And yet, until recently, it hadn't occurred to me to reflect on the way the figures are made—on their design, as distinct from their content.

This indifference to presentation has been dispelled by Edward R. Tufte's *The Visual Display of Quantitative Information*. The book was written with the aim of ensuring that people "will never view or create statistical graphics the same way again." By turns descriptive, proscriptive, and prescriptive, it contains some of the best examples of visualized data and, by way of warning, some of the worst. Tufte extracts rules of construction, combining aesthetic, scientific, and purely practical considerations to compile a kind of canon for designers of statistical displays. Finally, the lessons drawn from about two hundred examples are summarized in such maxims as "Above all else, show the data" and "Encourage the eye to compare different pieces of data." A persistent theme is that the more ink is used to represent actual measured data, and the less for auxiliary or decorative purposes, the better will be the graph.

Data graphics, according to Tufte, are the visual display of measured quantities by means of points, lines, numbers, symbols, words, shading, color, and coordinate systems. In this sense, maps are data graphics, and so are tables such as those that depict stock market prices in newspapers. The most common, though, is a series of points (or sometimes bars), perhaps connected by a curve, to indicate the dependence of one variable on another. If the graph's horizontal axis is time, the graph is called a time series; if the axis represents some other variable, it is simply called a graph (though it sometimes goes by the name bivariate scatterplot).

The most memorable feature of Tufte's book is his demonstration of the way that ugly, confusing, and misleading graphs can be turned into forceful instruments of data transmission. By rotating a graph's labels for easier reading and erasing a few redundant grid marks, by deftly lettering and tastefully shading, and by sensitively choosing format and size, Tufte makes silk purses out of sows' ears while teaching the reader to spot the visual junk and graphic lies that pollute our media. Further, he cautions that all the information a graph contains should be consequential—it should be about such issues as life and death and the universe. Beautiful graphics do not traffic with the trivial. What emerges from all this is a theory of data graphics that helps designers "to give visual access to the subtle and the difficult—that is, [to reveal] the complex."

Change but one word in that statement and it becomes the goal of those of us who write about science for the public: to give *verbal* access to the subtle and the difficult—that is, to reveal the complex. Whatever pictures we would portray are composed not of lines and color but of nouns and verbs. Are there rules for this endeavor, too, like those developed by Tufte for his? Is there excellence in this craft, and, if so, how is it constituted?

The standard for excellence in graphic display was set more than a century ago by what Tufte claims "may well be the best statistical graphic ever drawn": a map of Napoleon's Russian campaign of 1812 to 1813, created in 1861 by the French engineer Charles Joseph Minard. The map is well known; it can be found, for example, in some editions of Leo Tolstoy's *War and Peace.* What distinguishes Minard's version is that the thickness of the lines tracking the route to Moscow and back represents the dwindling strength of the French Army from one month to the next; the red line that starts at the western border of Russia is almost an inch wide, corresponding to 422,000 men. As it winds eastward, it becomes thinner, reaching Moscow with only a quarter of its former width— meaning that three quarters of the French troops were lost by that point. But it is the diminishing width of the black return line that tells the most dreadful chapter of the story. At the end of a harrowing winter retreat, the beleaguered army that staggered back across the Polish border numbered a mere ten thousand men.

By juxtaposing the broad red ribbon of the invasion with the meager black tread of the retreat, Minard delivers an emotional wallop. The graph compels the viewer to reflect on what Tolstoy called the melting away of the French Army at the uniform rate of a mathematical progression, punctuated by such catastrophes as the crossing of the Berezina River, where, according to the chart, twenty-two thousand soldiers died.

Whereas Minard used simple images to convey complex statistical information, the science writer strives to give simple verbal expression to measured data, to functional relationships among sets of data, and to theories, all of which, in their original form, are couched in the language of mathematics. The difficulty starts at the simplest level, with the description of numbers. Only a narrow range of numerical values can be grasped intuitively—those falling between, say, ten thousand (a number you can actually count to in three hours) and one one-thousandth (the thickness of a human hair when measured in inches). Since most of physics deals with numbers outside this range, the writer must translate them into intelligible word pictures.

In the realm of atoms and molecules, for example, objects are very tiny, but there are lots of them, so both small and large numbers abound. To establish a sense of comprehensible scale, a famous quantity, called Avogadro's number (after Count Amadeo Avogadro, of Turin, who introduced the idea in 1811), can be used. Roughly speaking, Avogadro's number represents the quantity of water molecules that can be held in a tablespoon, a number that is impossibly large by human standards—about 6.02×10^{23}.

What can writers do to convey a feeling for such a number? In one textbook, it is written out, 602 followed by twenty-one zeros, then described as six hundred two thousand million million million. This is indeed a translation into words, but one no more illuminating than Carl Sagan's phrase "billions and

billions of stars." In another textbook, the author—out of
frustration I suspect—resorted to visual imagery. After quoting
the value of Avogadro's number, the text refers the reader to
a figure that, in its entirety, consists of a large panel bearing
in fancy script the message "Avogadro's Number 6.02 $\times$
10^{23}." Tufte would not award high marks to that data graphic.

A much more successful attempt at conveying the immensity of Avogadro's number makes use of a theorem called Caesar's Last Breath. The theorem holds that with every breath
you take, you inhale a single molecule of the air that Julius
Caesar exhaled as he died. Behind the statement, of course,
lurks a qualification and several assumptions. First, the qualification: since it is based on an average, the theorem can be only
approximately true. If you take three or four breaths without
encountering one of Caesar's air molecules, don't be disap-
pointed; later you may inhale several at once. One of the
assumptions the theorem makes is that in the past two millennia no air molecules have been added to the atmosphere or
been taken out of circulation by becoming attached to other
molecules—which isn't true. Also taken for granted is that
Caesar's last breath has had a chance to mingle evenly throughout the atmosphere—which isn't quite realistic, either. But if
these assumptions *were* tenable, the theorem would be true
and could as well be applied to the dying gasp of Shakespeare
or Attila or Jesus.

The point of the parable is that, although the earth's atmosphere is vast, the number of molecules in a human breath is
also vast. If you were to measure the volume of the atmosphere
with a bottle having the same capacity as your lungs, the total
number of bottlesful would equal the number of molecules in
one breath. And both quantities amount to about one tenth of

Avogadro's number. This is probably the best verbal description that I have encountered of a large number.

Avogadro's number is unimaginably large, but even much smaller numbers are beyond our comprehension. Consider the approximately one trillion data graphics printed annually (an estimate derived from the introduction to Tufte's book). There is no way to represent a trillion graphically, so it is most often put either into words (a thousand billion) or into mathematical notation (10^{12}). The most common word picture of such a number, dear to newspaper reporters, is expressed in terms of distance; for example, if a trillion copies of Minard's map were laid end to end, they would reach the sun. But so what? I can imagine neither that distance nor the trail that papers it.

A better story might go as follows: Imagine a thousand-page book of graphics, about as thick as *War and Peace,* in which each page bears a single picture. Then imagine a train with all the seats filled and every passenger leafing, at leisure, through a copy of this book. The train is long—so long that, standing at the side of the tracks, you would have to wait six hours for all of its cars, traveling at full speed, to pass. Now, if this train ran once a day for a year and, on each trip, each passenger thumbed through the entire fat book, a trillion visual impressions of data graphics would have been registered. (I am indebted for this image to the physicist Phillip Morrison and his colleagues, of the Boston Study Group, who once imagined a train of such length filled with TNT to illustrate the enormity of one megaton.)

Caesar's breath and Morrison's train succeed as verbal descriptions of large numbers because both refer to common human experiences. Millions of millions of millions is not a quantity that is accessible to the human senses, and neither is the distance from the earth to the sun. But breathing, reading, and riding crowded trains, by contrast, are familiar activities and are therefore more comprehensible.

Even more difficult to convey than numbers are complex relationships among variables. The graph of a linear function, the simplest kind of relationship, is a straight line, but it is often verbally described as a proportion: if x doubles whenever y doubles, then x and y are linearly related—in the same way that the number of eggs you buy is linked to the total price paid for them. Linear relationships, sometimes known as arithmetic progressions, don't pose problems for the graphic designer or the writer.

More challenging are geometric progressions, such as those encountered in the accrual of interest, in population growth, and in atomic chain reactions. There are many verbal illustrations of the rapidity of geometric growth, involving everything from dollars to rabbits, but one of the best is also probably the oldest. It is an ancient legend that might be entitled "The Royal Chessboard." One account, in James R. Newman's *The World of Mathematics,* reads:

> The Grand Vizier Sissa Ben Dahir was granted a boon for having invented chess for the Indian King, Shirhâm. Since this game is played on a board with 64 squares, Sissa addressed the king: "Majesty, give me a grain of wheat to place on the first square, and two grains of wheat to place on the second square, and four grains of wheat to place on the third, and eight grains of wheat to place on the fourth, and so, Oh, King, let me cover each of the 64 squares of the board." "And is that all you wish, Sissa, you fool," exclaimed the astonished king.

Evidently, the king did not understand geometric progressions. So the wish was granted, and Shirhâm ordered wheat to be brought from his stores. By the fortieth square, a million million grains had to be hauled up, and long before the last square was reached, the kingdom's entire wheat supply was exhausted. In fact, Sissa's demand was just one grain shy of 2^{64} grains, or 18,446,744,073,709,551,615—enough to cover the surface of the earth to the depth of an inch. Applied to atomic fission, in which each rupture of an atom's nucleus triggers the rupture of approximately two others, the tale readily explains how a single nuclear breakup can initiate a process that will quickly grow to explosive proportions.

At the highest level of mathematical sophistication, even beyond numbers and relationships, are theories. The device for putting them into words is analogy, the very essence of scientific popularization. Physicists are comfortable with analogy because in some sense their whole enterprise, the modeling of nature by means of mathematical constructs, is analogical. Thus, analogies abound not only in popularizations but in the professional literature as well. And among theories, too, there are classic word pictures.

Few theories command as much public curiosity as Einstein's general theory of relativity. It is not surprising, then, that the best analogy in all of physics concerns relativity, or that it was invented by the master himself. Known as Einstein's Elevator, it is a verbal device that presents the crux of the general theory of relativity, which, in its mathematical formulation, is completely inaccessible to most laymen.

A cornerstone of relativity theory is the principle of equivalence, the assertion that gravitational fields are equivalent to accelerated frames of reference. This statement, obscure in itself, was illustrated by Einstein in the following manner: Imagine, on the surface of the earth, a large closed box within which a man performs simple experiments—dropping an apple and timing the swing of a pendulum. Imagine an identical box, far out in space, beyond the gravitational reach of the earth, the sun, or any other star, in which a woman performs the same experiments as the man does on the earth. Now a rope is attached to the top of the box in space, and an unseen agency pulls it upward with an acceleration which is numerically equal to the gravitational acceleration on earth. Thus the box in space resembles an ascending elevator.

The principle of equivalence asserts that the two environments are indistinguishable—that neither the man nor the woman can tell, without peeking outside, whether his or her own box is resting on Earth or accelerating in space. If the man drops an apple, it will hit the floor after a short interval of time. The woman's apple, when released, hovers in space, but the floor of her box rises to hit it after the same interval, so she, too, thinks that the apple has dropped to the floor. The effect of gravity is simulated by the acceleration of the box which constitutes her frame of reference.

Einstein's elevator lucidly explains the strange central idea of general relativity, but it can do even more: it is powerful enough to predict new phenomena. Suppose the man affixes a flashlight to a wall, three feet from the floor, and aims it horizontally across his box. Where will the beam strike the opposite wall? Three feet from the floor? Before answering, switch to the woman and have her perform the same experiment. As the beam of light speeds across her box, the floor rises—not much, because it doesn't take long for light to cross the short distance from wall to wall, but in principle the floor does rise a fraction of an inch. So her beam hits the opposite wall *less* than three feet from the floor. Back to the man on earth. If the principle of equivalence is correct, his experimental results should be the same as the woman's. He, too, should find the beam hitting the opposite wall less than three feet from the floor, allowing only one conclusion: the beam is pulled down by gravity. In other words, the principle of equivalence predicts that light is deflected by gravity.

This remarkable prediction cannot, in reality, be tested in a box or a room, because the distance that light would drop is too small to be measured. The theory is amenable to experiment on an astronomical scale, however, and in 1919 it was shown that the sun's gravitational field does indeed deflect starlight, thereby verifying the prediction and ensuring Einstein's immortality. To be sure, the mathematical details are much more complicated than the elevator experiment would suggest, but the story is, after all, only an analogy, not a theory.

Stories such as "Caesar's Last Breath," "The Royal Chessboard," and "Einstein's Elevator" follow a pattern that suggests the formulation of rules for analogizing. First, state the problem. It should be important, of universal concern, and genuinely incomprehensible, for good analogies should not traffic with the frivolous or the obvious. Next, introduce a familiar object or everyday activity that is instantly recognizable; describing one mystery in terms of another is a waste of time. If, in the third step, you can lead a reader to see the relationship between the incomprehensible concept and its simple analogue, you have succeeded. But good analogies go one step further. By logical extension, they lead to something beyond what they were supposed to explain, to a prediction or an unexpected discovery, to something so surprising that it fixes the story in the reader's mind. When that happens, a precious instance of real learning has occurred.

A systematic analysis of the art of creating analogies might lead to a formula for the verbal description of quantitative concepts. But, like Tufte's manual of style for graphic design, a set of such rules would be, at best, suggestive. It might help to eliminate much that is useless or obfuscating in attempts to popularize science, and might guide science writers, but it could not guarantee memorable analogies. Those come along rarely. Minard's map and Einstein's elevator are unique products of imagination and inspiration, and should be appreciated as minor works of art. There can be no fixed prescription for producing them. The virtuosos of human expression do not so much abide by rules as embody them. As the painter Barnett Newman once said, "Aesthetics is for the artist like ornithology is for the birds."

13.

The Inverse Problem

When a man knows he is to be hanged in a fortnight, observed Samuel Johnson, it concentrates his mind wonderfully. The imminence of brain surgery has the same effect. One day many years ago, lying on a hard pallet in a tiny room in Massachusetts General Hospital, in Boston, with my head inside a CAT scanner, I was worried that the tumor in my brain was malignant, and frightened of the surgery scheduled for the next day. But I hadn't been anesthetized and wasn't in pain, so I was able to concentrate wonderfully on what was happening around me.

I had read about the six-foot steel cube into which my head had been placed. The clanging I heard and the tremors of movement I felt were the machine's X-ray apparatus repositioning itself as it slowly circled my skull, adjusting the direction from which it beamed radiation toward my brain. The many images it produced were being fed into a computer (hence the term *CAT*, for computerized axial tomography). A radiologist presented with these various flat, partial views would not have been able to make much sense of them. But the computer was quickly performing the mathematics needed to correlate the images into a coherent three-dimensional picture of my brain that would allow a surgeon to peer into my skull before removing the tumor (which proved to be benign).

As the machine went noisily about its work, it occurred to me that the CAT scanner's operation is a model for the way we solve scientific problems, and perhaps even for the way we perceive and reason. The machine is just one solution to what physicists call the inverse problem, a problem common to such disparate endeavors as prospecting for buried minerals, charting the earth's core, analyzing the atomic constituents of crystals, dissecting the atoms themselves—indeed, any effort to map and measure an object that cannot be seen. Regardless of the materials involved, the solution is the same: A stream of subatomic particles (or, mutatis mutandis, of radiation) is beamed toward the unknown target. Either the projectiles are absorbed or they bounce off, and by carefully correlating the results of many encounters, one can detect the pattern and deduce from it the shape of the target. It is this backward tracing of the particle's path that gives the inverse method its name. Of course, tracing hundreds of projectiles requires extensive calculations that only a fast digital computer can manage, which is why most practical uses for the inverse approach are relatively recent. It was only in the late 1980's that the field's first scholarly journal, *Inverse Problems,* (published by the Institute of Physics in London), was started.

Yet it seems strange to think of the solution of inverse problems as a new approach in physics. The task of deducing reality from the observable data is the quintessential procedure of empirical science. If the inverse method is a novelty, then what have physicists been doing all these centuries? In a word, they have been guessing. Although they have begun with observations and proceeded to discover the unknown, they have not done so systematically. Rather, they have been jumping to conclusions before solving a single equation.

Consider the contribution of the physicist Ernest Rutherford to the discovery of the structure of the atom. In 1911, Rutherford was supervising an experiment in which alpha particles, a form of radiation, were beamed at a very thin gold foil. He expected the beam to emerge on the other side of the foil only slightly deflected because, like most physicists of the time, he thought that mass is evenly distributed within the atom, just as it is in a marble. But most of the particles were not deflected at all, and a few veered sharply off course.

Guided only by hunches, Rutherford guessed that the atom must consist of a tiny dense central core surrounded by a tenuous cloud of electrons. Most of the alpha particles had torn through the electron cloud, but those that hit the atom's center had been deflected. Rutherford worked out the equations that predicted how alpha particles should bounce off his model of the atom and, seeing that these calculations matched exactly the scattering he had observed, concluded that he had discovered the atomic nucleus.

This direct approach (making a guess and seeing how it fits), as opposed to the inverse method, is a time-honored way of answering questions in science. Guided by intuition, imagination, and inspiration, scientists guess at hidden shapes, then calculate forward to predict how the projectiles should scatter if the guess was correct. If the two predictions agree, the guess is considered good. If not, another guess is made, and then another, until observation matches calculation. The trial-and-error approach is fundamental to scientific thinking, as the late Richard Feynman explained in *The Character of Physical Law:*

In general, we look for a new law by the following process. First we guess it. Then we compute the consequences of the guess to see what would be implied if this law that we guessed is right. Then we compare the result of the computation to nature, with experiment or experience, compare it directly with observation, to see if it works. If it disagrees with experiment it is wrong. In that simple statement is the key to science.

The contrast between the direct and the inverse methods is poignantly illustrated by the story of the discovery of the structure of dioxyribonucleic acid. The crucial clues were the symmetrical but seemingly meaningless patterns of dots and streaks that appeared on X-ray photographs of crystallized DNA. James D. Watson, in *The Double Helix*, the highly personal account of how he and Francis Crick came to find the solution, praised Linus Pauling for demonstrating the advantage of a good guess over the painstaking analysis of those photographs. In 1951, Pauling had shown that many proteins contain parts that are helical in form. "The helix had not been found," Watson wrote, "by only staring at X-ray pictures. . . . In place of pencil and paper, the main working tools were a set of molecular models superficially resembling the toys of preschool children." Watson and Crick, like Pauling, had built a Tinkertoy-like model first, and only then did they compare their predictions with the photographs.

The Double Helix also tells the sad story of Rosalind Franklin, who produced the wonderfully sharp photographs of DNA that were essential to the discovery of its structure, but who failed to interpret them correctly because she disdained the use of models based on mere hunches. Franklin attempted to deduce the structure of the DNA molecule directly from her photographs, and was undone by her unwillingness to turn the inverse problem around.

Of course, Watson and Crick, like Rutherford and Pauling, had guessed correctly, and that made all the difference. But how did they guess? The best theorists, it would seem, get through the boring business of calculating forward faster than anybody else. By accelerating the process of trial and error, they have a better chance of hitting the right solution.

And yet if that were the whole story, there would be no distinction between the idiot savant and the creative genius. Blind guessing, Feynman wrote, "is a dumb man's job," and he went on to trace the ways in which inspired hunches have come to him, and to provide guidelines for eliminating wrong assumptions. But the real nature of scientific creativity, and of genius, remains elusive.

Perhaps the inverse approach has something to do with it. In *A Feeling for the Organism,* Evelyn Fox Keller's biography of the geneticist Barbara McClintock, who won a Nobel Prize in 1983, McClintock described being told about the discovery of corn plants near her laboratory with traits that seemed to defy the rules of heredity. She thought about the mystery, and suddenly the answer came to her. So great was her excitement and surprise, she said, that she ran out to the field and actually shouted "Eureka!" "Now, why," she asked, "did I know, without having done a thing on paper?"

The answer, for McClintock, lies in her long experience. She has a profound feeling for the organism, and this is not simply a matter of knowing more about her subjects than anyone else. She believes that, in all the years she has worked with corn plants, her mind has been analyzing far more information than she was aware of. (She even goes so far as to compare this subconscious process with the work of a computer.) Perhaps, then, a scientist's sudden flash of intuition is actually the solution of an inverse problem of years' standing, patiently worked through by the subconscious. Perhaps, too, inverse solutions are not so novel and Thomas Edison came closer to the truth than he realized when he said, "Genius is one per cent inspiration and ninety-nine per cent perspiration."

14.

Impossible Crystals

"I do not know what I may appear to the world; but to myself I seem to have been only like a boy, playing on the seashore." Thus Sir Isaac Newton, the patriarch of modern physics, defined his life's work. And in so doing he revealed a truth about his vocation that is rarely apparent to nonscientists. Too often the playful child in the center of the enterprise is hidden by a wall of abstruse theory and an impenetrable welter of technology. But occasionally we see a little face peeking out from behind the mask of profundity, and from such rare glimpses we gain better insight into the true nature of science than from volumes of erudite explanation.

The vital role of scientific playfulness is vividly illustrated by the story behind the discovery of a new class of materials known as quasicrystals. Quasicrystals are three-dimensional structures, but their antecedents exist in two dimensions, in the plane. The story begins in January 1977, when Martin Gardner devoted his Mathematical Games column in *Scientific American* to the question of how to cover a plane with tiles. It's a problem as ancient as Greek mosaics, yet Gardner's essay sparked a flurry of research that brought tiling to the forefront of modern physics.

The mathematical analysis of tiling begins with the observation that a plane—a bathroom floor, for example—can be covered without gaps by tiles in the shape of rectangles, triangles, or hexagons, but not by circles or stars or even, significantly, regular pentagons. No matter how you try to join pentagonal tiles, they will always leave gaps. To convince yourself of that, just cut out a pile of identical pentagons and play with them on your desk. You will soon develop a distinct antipathy toward the five-sided monsters.

A regular pentagon has fivefold symmetry, which means that if you rotate the pentagon about its center, it looks the same after every one fifth of a rotation. Similarly a square has fourfold symmetry, a hexagon sixfold, and so on. Any shape that tiles a plane can impart its symmetry to the whole tiling pattern: you can rotate a hexagonal tiling about the center of any hexagon, for example, and see that the whole pattern has sixfold symmetry. If you could use a pentagon to tile a plane, then the tiling could exhibit fivefold symmetry. But, of course, pentagons are prohibited.

Using two tile shapes instead of one, however, Gardner reproduced intriguing tiling patterns that did indeed display fivefold symmetry: whole regions could be rotated so that the arrangement of tiles within looked the same every one fifth of a rotation. One pattern, for example, contained groups of tiles that looked like five-pointed stars. It was as if pentagons were defining the rules without actually being present.

Gardner learned how to construct these puzzling tilings from Roger Penrose, a British mathematical physicist. Penrose has a knack for playing tricks with geometry. In his youth he and his father drew an impossible object, the "Penrose staircase," which spirals round and round without getting higher or lower. The Dutch artist M. C. Escher used their mind-boggling concept in the famous lithograph *Ascending and Descending*, which shows a line of men going both up and down the stairs simultaneously.

"Penrose tiles" are equally intriguing. Neither of the two basic shapes is a pentagon, and they do not combine to form a pentagon. But the shapes are like mischievous pentagonal offspring: they have angles and proportions that can be found in a pentagon and its diagonals, and when assembled on a plane the two proudly display the fivefold symmetry of their parent.

The simplest Penrose tiling uses two diamond shapes, one fat and the other skinny. The fat shape has an interior angle of 108 degrees, the interior angle of a regular pentagon. The skinny shape has an interior angle of 36 degrees, an angle formed by the pentagon's diagonals. While all the tile edges are of equal length, the ratio of the area of the fat tiles to that of the skinny tiles is $(1 + \sqrt{5})/2$, which equals approximately 1.618. This happens to be the ratio of the length of a diagonal to that of a side of a regular pentagon. It is also the famous "golden ratio," a measure revered as a standard of harmony by both the ancient Greeks and a legion of Renaissance painters and architects. As we shall see, the golden ratio and pentagonal symmetry are embedded in the design of Penrose tilings in many wondrous ways.

Any Penrose tiling can be constructed in an infinite variety of patterns. Every variation is nonperiodic, and therein lies its allure. Unlike the individual bricks in a wall or the pickets in a fence, no group of one or more tiles can be repeated indefinitely to generate the whole pattern. At first glance Penrose tilings may look periodic—groups of tiles do form such repeating motifs as five-pointed stars—but a more careful look reveals that the spacing between these motifs is irregular, and some are rotated with respect to others.

Naturally, when researchers saw these patterns balance teasingly between order and chaos, they were drawn to them like children to a brand-new toy. Over the next half-dozen years many Penrose tilings were generalized to three dimensions, using solid polyhedrons that fill space without gaps. Like their counterparts in the plane, the three-dimensional tilings were also nonperiodic.

One of the enchanted players was Paul Steinhardt, a physicist at the University of Pennsylvania, who is well aware of the research value of playthings. Steinhardt's office is filled with toys. Scattered among the books and computers that are the standard trappings of the scholar's craft is every conceivable kind of model, from the crudest cardboard cutouts untidily held together with tape to expensive computer graphics. Anything at hand is pressed into service: coat hangers, foam balls, dice from the game Dungeons and Dragons, acetate sheets, Tinkertoy pieces, toothpicks, construction paper. Steinhardt is a natural victim for the kind of game that can be played with three-dimensional nonperiodic tilings.

Indeed, in 1984 he and one of his graduate students, Dov Levine, became so caught up in the game that they took the analysis one step further: they programmed a computer to calculate the diffraction patterns these theoretical structures would produce if the building blocks were real atoms instead of imaginary tiles.

Diffraction patterns are the windows physicists use to peer inside materials. When beams of electrons or X rays pass through solid materials, they are diffracted, or scattered, by the atoms inside. The diffracted beams can be photographed head-on, and the images they form on the film reflect the atomic architecture of the solid object they've just passed through. By themselves diffraction patterns are not much to look at. They consist of mysterious arrangements of dots and streaks that bear little resemblance to the solids they portray. But to the initiated they are as recognizable as family snapshots.

The most distinct diffraction patterns contain sharp, isolated dots. These are the portraits of crystals, and they owe their clearly defined spots to the periodicity of the underlying structure. When the beams hit the atoms in a crystal, they scatter in all directions; but in a few preferred directions, depending on the arrangement of atoms, the diffracted beams reinforce one another, producing bright spots on the film. A crystal is a little like an orchard planted in a rigid geometric grid. Most lines of sight are blocked by trees, but in a few directions you can see right through to the other side.

In another class of diffraction patterns the dots are either spread out into fuzzy rings or altogether absent. These are the images formed by glassy materials. Glasses, in contrast to crystals, consist of atoms or molecules stuck together randomly; they're more like random forests than well-planned orchards. Because they offer no preferred directions for diffraction, the patterns they produce contain no sharp dots.

Until the discovery of quasicrystals, it was thought that there were only these two classes of solid materials, corresponding to these two types of diffraction patterns. If the pattern contained sharp dots, the material was a crystal; if the dots were fuzzy or absent, it was a glass. Every pure solid in nature, from gemstones to metals to DNA, was either crystalline or glassy.

Levine and Steinhardt called that neat scheme into question when they aimed a simulated X-ray beam at one of their imaginary solids. The computed diffraction pattern contained a surprise: unmistakable sharp points. Since the atomic arrangement of their solid was nonperiodic, it should have produced the fuzzy diffraction pattern characteristic of glassy substances.

This contradictory result required an explanation, of course, and to understand what was going on, the two physicists went back to the source of their computer model: the two-dimensional Penrose tiling. They also consulted Robert Ammann, a recreational mathematician. Ammann's work led them to the discovery that the spacing between the tiles was neither periodic nor random but something in between, an order called quasi-periodic.

The distance between tiles, Ammann found, is one of two lengths, either a longer length, *a*, or a shorter length, *b*. The ratio of the longer to the shorter is the golden ratio, and the two lengths succeed each other in a predictable, fixed order— an infinite series known as the Fibonacci sequence. Leonardo Fibonacci, a thirteenth-century mathematician, defined the rules for producing this sequence when he considered the idealized propagation of rabbits. The first rule: Start with one adult rabbit, *a*, and assume that at the end of every year each adult has a baby, *b*, which you record right after its parent. Second rule: Every baby grows into an adult the year after it is born. (Fibonacci, a realist, made each letter stand for a *pair* of rabbits. Steinhardt and many others simplify by pretending that a single parent can have a baby, and we follow the simpler description here.)

To get the first term of the sequence, you start with one adult: *a*. In the second year the adult has a baby, and the sequence goes to two terms: *ab*. In the third year the original adult has another baby, which is recorded after the adult, and the first baby grows up, to give you *aba*. In the fourth year you add a baby after each adult, and change the baby that was already there to an adult, getting *abaab*. If you keep going year after year, you get *abaababa, abaababaabaab,* and so forth. Another way to generate the sequence is to add the sequences of the two previous years, writing last year's sequence first. And so the sequence does not change from year to year; it just grows longer. If you write down the total number of rabbits in each year (1 in year one, 2 in year two, 3 in year three, 5 in year four, and so on), you get a string of integers that make up the famous Fibonacci sequence (1, 2, 3, 5, 8, 13, 21, 34, 55, 89 . . .), in which each term is the sum of the previous two. And again the golden ratio rears its beautiful head. As the series progresses the ratio of any two successive terms approaches 1.618.

Obviously Fibonacci was onto something. Nearly eight centuries ago he invented a sort of one-dimensional Penrose pattern, a sequence that while not periodic is not random either, for there is a perfectly rigorous prescription for predicting what the next member of the sequence will be. Thanks to the recreational mathematics of the Middle Ages, Steinhardt and Levine were onto something, too. They finally had an explanation for the diffraction pattern of their imaginary solid. Their discovery amounted to the demonstration that, contrary to established belief, periodicity in three dimensions was not necessary for producing diffraction spots—quasi periodicity was quite sufficient.

But their quasi-periodic solid had an awkward feature: the dots in the diffraction pattern were arranged with fivefold symmetry. To traditional crystallographers, such a pattern is simply unacceptable. It is a fundamental tenet of their science that crystals cannot produce diffraction patterns with fivefold symmetry because the underlying arrangement of atoms cannot have pentagonal symmetry—any more than bathroom floors can be tiled with pentagons.

So whatever Steinhardt's imaginary solids were, they were not crystals. But the discrete spots in the diffraction patterns showed they were not glasses either. Their underlying structure combined properties of crystals with those of glasses; the theoretical substances were like mammals that lay eggs, the platypuses of physics. Steinhardt decided to call them quasicrystals, and added them to the growing list of amusing ideas that spring from Penrose's mathematical recreation.

But then the incredible happened. In the fall of 1984, while Steinhardt was on leave at the IBM research center in Yorktown Heights, New York, a colleague, Harvard physicist David Nelson, came into the office with exciting news. He put on the table a small copy of a diffraction image made with a real alloy of aluminum and manganese. Nelson first explained that a team of researchers at the National Bureau of Standards had made the picture, then pointed out the unusual appearance of the pattern of dots: an obvious fivefold symmetry.

Steinhardt's pulse quickened. The picture looked amazingly similar to a computer simulation he and Levine had produced and not yet published. Fortunately, Levine happened to be visiting from Philadelphia that day, and immediately the three scientists, as excited as boys playing on the seashore, set to work measuring the spacing between dots on an enlargement of the real photograph and comparing the results with the computer printout. Steinhardt recalls that he knew what the answer would be even before the measurements confirmed it. The two pictures agreed with each other.

The moment of truth, in science, comes when theory confronts experimental evidence. Agreement between the two is the ultimate arbiter of validity. Nothing else matters. The comparison of data with calculations usually proceeds in bits and pieces, and truth emerges gradually from the confusion that surrounds all creative effort. But when the moment of truth arrives in an unexpected flash, as it did that day at IBM, it illuminates and energizes the scientific enterprise for years to come.

Thus a new field of solid-state physics, the science of quasi-crystals, was born. It grew up quickly. In short order more than a hundred alloys with fivefold symmetry were discovered; sevenfold, ninefold, elevenfold, and other previously forbidden symmetries proved to be possible; scholarly symposia were convened, and fat monographs published.

But in the closet there lurked a skeleton—a potentially fatal flaw in the whole scheme. While researchers were beginning to understand the architecture of two- and three-dimensional quasi-periodic tilings, no one could think of a mechanism by which millions upon millions of real atoms could arrange themselves spontaneously in those intricate patterns.

Anyone who tries to assemble Penrose tilings quickly realizes that it's not easy. You have to think ahead and keep the whole pattern in mind when adding a tile; otherwise there is trouble. If you make a mistake, you have to undo a lot of work that has gone before. The problem is that while there are local rules, or instructions for fitting a tile into a particular niche, these rules are not sufficient to build the entire pattern. It seems necessary to augment them with global rules that force you to plan ahead and check the configuration of tiles at far distant points. And between 1984 and 1988 the conviction grew that perfect quasi-periodic tilings could not be constructed with local rules alone.

Local rules for adding tiles are analogous to forces that attract and hold new atoms to the surface of a growing quasi-crystal; they are plausible ingredients in the growth mechanism. Global rules are not. The atoms on a growing surface do not plan ahead, and they do not check the orientation of distant surfaces; they respond only to the interatomic sticking force, which is electrical in origin, of their immediate neighbors. If quasi-periodic patterns could be constructed only with the help of global rules, they could not be assembled by real atoms in real alloys, and quasicrystals could not exist in nature.

The problem was so serious that researchers began shifting their attention to more conventional explanations of the observed diffraction patterns. The two-time Nobel laureate Linus Pauling, for example, championed an arrangement of ordinary crystals called twinning. Twinned crystals grow from separate origins and penetrate each other at odd angles, such as 72 degrees. This might produce a diffraction pattern with spurious fivefold symmetry even though the underlying structure was conventional. Other researchers, including Steinhardt himself, studied glassy structures with tiny embedded crystalline fragments to see if they would be capable of causing a diffraction pattern with spots almost as sharp as the dots from crystals.

But then, in 1988, playfulness paid off once more. George Onoda, an IBM ceramics expert, started toying with about two hundred Penrose tiles. Unconvinced by the claims that he wasn't supposed to be able to do it, he learned how to assemble flawless tilings of any size he wished using strictly local rules. "I approached it as a puzzle," he says, "as a challenge to try to prove the naysayers wrong."

Onoda showed Steinhardt his procedures, and the two of them fiddled around with the tiles for a couple of hours. Steinhardt simplified Onoda's insights to a set of rules that force the vertices of the tiles into one of the eight possible combinations found in a perfect Penrose tiling. With the help of two other researchers, he then hammered out a mathematical proof, corroborated by a computer simulation of a million tiles, that quasi-periodic structures can indeed grow naturally—at least in two dimensions.

Following these rules, you can build a Penrose tiling by adding tiles to a growing boundary. The rules specify which type of vacancy to fill first and to choose randomly if there is more than one equivalent vacancy, which of the two tile shapes you should use in every case, and which way it should be turned. You don't have to pay attention to any distant part of the pattern to assemble a tiling—any more than an atom has to know what's going on somewhere else before it decides which way to turn and attach to its neighbors.

For a complete theory of quasicrystals the local rules must be generalized to three dimensions, and they must be shown to correspond to actual atomic forces. Neither of these tasks has been achieved yet, but Steinhardt, for one, believes they will be.

In the meantime, experimentalists have been busy. They continue to report bigger, more perfect quasicrystals and are diligently measuring their physical properties. No one knows what to expect, for none of their vast experience with crystals and glasses permits them to make confident predictions about quasicrystals. Quasi-crystalline alloys, because of the intricate interlocking of their constituents, might turn out to be harder than crystals and might therefore be used as replacements for industrial diamonds. Or they might end up at the heart of novel electronic devices as yet undreamed of. Who knows? Researchers are going to have to play around with them a bit to see what they can do.

15.

The Elusive Monopole

Particle physics lost a member of the family the other day. To be sure, this particular individual was a bit of a loner, an interloper like an unexpected uncle from the antipodes, but perhaps for that very reason its loss was the more keenly felt. What happened was that the only observation of a magnetic monopole, recorded just before two o'clock on the afternoon of St. Valentine's day in 1982, was retracted. And the world of particle physics is the poorer for it.

Magnetic monopoles are isolated poles, labeled either north or south, and apart from that one spectacular exception, none has ever been found. Unlike electrical charges, which also come in two varieties (called positive and negative), and which are usually observed in isolation, magnetic poles seem to occur only in pairs called dipoles. Magnets have two ends, compass needles two tips, and planets two polar regions. Even elementary particles, such as electrons and protons, are endowed with minuscule magnetic dipoles. Nature seems to abhor magnetic monopoles.

Every attempt to produce a monopole artificially has been frustrated. In the great magnetic monograph *De Magnete*, which was published in the year 1600 and is sometimes called the first treatise of modern science, Queen Elizabeth's physician William Gilbert described what happens when you cut a magnet in two: instead of a north pole and a south pole, as you might expect, you get two magnets, each complete with its normal complement of two poles. Today we know how to make electromagnets out of coiled current-carrying wires, but every configuration we try, no matter how ingenious, displays the familiar pair of poles. North and south are as inseparable as the two sides of a coin.

In theory, magnetic monopoles seem to make eminent sense. When the Scottish physicist James Clerk Maxwell created the unified theory of electricity and magnetism in the 1860s, he found many wonderful parallels between the two. In fact, his equations exhibit a high degree of symmetry between electric and magnetic fields. One can exchange their mathematical symbols, and almost, but not quite, get back the original formulas. The pattern is spoiled only by the presence of a term for electrical charges, and the absence of its magnetic counterpart.

The broken symmetry of Maxwell's equations fairly cries out for restitution by the introduction of magnetic monopoles, and thus justifies the search for such objects. Many of the elementary particles known today were discovered in the course of such attempts to fill out pleasing but incomplete theoretical patterns. In spite of their lack of symmetry, Maxwell's equations are so well established that their predictions about the behavior of magnetic monopoles, should they exist, can be trusted. Paradoxical as it may sound, except for the mystery of their nonexistence, monopoles are quite well understood.

In fact, they can even be mocked up experimentally. Imagine a very long, thin, flexible magnet, made, say, of a large number of tiny bar magnets pushed end to end into a mile-long hose as thin as an intravenous feeding tube. The effects of all the poles are canceled by those of the opposite poles touching them, except at the two ends. Since those ends are so far apart, they don't perceptibly affect each other: each one represents a monopole, and behaves like one. Only that awkward tube, an unavoidable umbilical cord between the two poles, spoils the simulation. As long as that appendage doesn't get in the way, the behavior of a monopole can be explored in this way.

Considerations such as these have been part of the classical canon of physics for over a century. During the last decade, however, interest in magnetic monopoles has reached new heights for both theoretical and experimental reasons. On the theoretical side, the most comprehensive of the schemes for describing all particles and forces, the so-called grand unified theories, unequivocally predict the existence of monopoles. Experimentalists, for their part, have invented new detectors, based on the phenomenon of superconductivity, that are ideally suited for finding them. Physicists were ready for monopoles in the early 1980's, and greeted the St. Valentine's Day discovery by Blas Cabrera at Stanford University with excitement and delight.

Since then, the experiment has been repeated by laboratories throughout the world—without success. Cabrera himself improved his apparatus several times. Finally, in February of 1990, reporting on the negative result of an experiment that was two thousand times more sensitive than the first one, Cabrera and his colleagues advised that the original data "should be discarded." The magnetic monopole became a nonparticle by edict of its discoverer.

The disclaimer raises interesting questions. Nobody, not even Cabrera himself, has found an error in the original experiment. Nor have new theoretical arguments cast doubt on it. The retraction is a judgment call, based on the overwhelming negative evidence of the intervening eight years. But does that really invalidate the first discovery? It is possible to destroy material evidence, but can you "undo" an experiment? Can you unpublish a poem?

Until 1931 the magnetic monopole was not much more than a curious mythical beast, a unicorn among elementary particles. But then respectability was conferred upon it by the great British physicist Paul Dirac, who won the Nobel Prize for bringing Einstein's theory of relativity to bear on the problem of atomic structure. Dirac asked how monopoles would fit into quantum theory, the fundamental description of the structure of matter. His answer turned out to be unexpected, and utterly tantalizing with respect to the question of the existence of magnetic monopoles.

Dirac considered a hypothetical heavy monopole, sitting at rest—say, on his desk. Somewhere far overhead a charged particle passes by. It is well known that a magnetic field will deflect a moving charge—for example, you can distort the picture on your TV screen by bringing a magnet up to it. The path of the charge flying over Dirac's head would therefore be bent slightly by the magnetic field of the monopole on his desk. When Dirac thought carefully about the nature of this deflection, he realized that the laws of quantum mechanics allow only certain minute, discrete changes in path—quantum leaps—which are expressed by integers. Since the amount of deflection obviously depends on the strength of the monopole and the magnitude of the charge, he was able to deduce a formula that can be written in the appropriate units in a deceptively simple way: the strength of the monopole times the magnitude of the charge equals an integer.

This equation, called the Dirac quantization condition, can be derived in a number of different ways, and always comes out the same. Considering that it describes a nonexistent object, it is a well-established, robust fact, and its consequences are as sweeping as they are astonishing.

The strength of the putative monopole on Dirac's desk is not important—it is just some fixed number: call it x. Once that is given, every charge in the universe must be able to satisfy Dirac's quantization condition, meaning that multiplying it by x must yield a whole number. The smallest charge that can possibly exist and obey the required condition is equal to one divided by x. The next one is two over x, then three over x, and so on. Charges, in other words, can exist only in integer multiples of the "quantum of charge," which equals the reciprocal of x. The quantum of charge is the smallest coin in the realm of electricity.

The quantization of electric charge is not a new idea. In fact, it is an established law. The only charges that have been found to occur in nature, among dozens of types of particles and trillions of individuals, are integer multiples of 1.602177×10^{-19} in units of electrical charge, which are called coulombs. Thus, for example, all electrons have exactly the same charge, and protons carry an equal and opposite charge. Without these precise conditions, the world would be a very different place from what it is. What Dirac had achieved with his thought experiment of 1931 was to provide an explanation for the universal phenomenon of charge quantization. In order to understand it, all you need to do is to accept the laws of quantum mechanics, and then find one monopole somewhere in the universe.

The ingenuity of the argument has inspired physicists to devote considerable amounts of time and effort to the search for monopoles. In 1981, half a century after he proposed the idea, and a year before Cabrera's discovery, Dirac himself, with characteristic modesty, commented: "From the theoretical point of view one would think that monopoles should exist, because of the prettiness of the mathematics." But, he added prudently ". . . pretty mathematics by itself is not an adequate reason for nature to have made use of a theory."

As if Dirac's explanation of charge quantization were not enough, modern grand unified theories, or GUTs, which attempt to systematize not only the elementary-particle zoo, but also its origins in cosmology, predict the existence of magnetic monopoles. The so-called GUT monopoles would have exotic properties, of which the most spectacular is their mass. Whereas ordinary elementary particles, like protons, weigh so little that a billion billion are required to make up a speck of dust, each GUT monopole by itself would weigh as much as that speck, which would make them macroscopically observable elementary particles, more massive than microbes, as large in diameter as some atoms. Their magnetic strength would be the reciprocal of the charge of an electron, and their magnetic field enormous. Since they are so heavy, they would tend to move slowly, lumbering through space like dinosaurs, ancient relics of the big bang.

Since GUT theories are theoretically appealing, their characteristic monopoles have been sought far and wide. Could they be hiding deep in the earth? Dirt from mines has been examined, with no success. On the moon? Moon dust has been analyzed. Could they be made by particle accelerators? By nuclear reactors? No, and no again.

In the end, theorists turned their inventiveness to explaining why monopoles are *not* abundant in the universe, and in due time found a way to modify the theory of cosmic evolution in such a way as to account for their absence. However, it is in the nature of the physics enterprise that while experimentalists take inspiration from the speculations of theorists, they are not necessarily persuaded by every theoretical argument. Every proposition must be tested. So the search for monopoles continues.

One place to look is in interplanetary space. It turns out that there is already evidence for the sparsity of monopoles in that region. Monopoles are accelerated by magnetic fields the way electrons are accelerated by electric fields. In the process they sop up energy from the magnetic field, and thus deplete it. So the very existence of a strong interplanetary magnetic field argues against the existence of a great abundance of monopoles; if they were out there, they would have used up the field long ago. Thus, the known strength of the interplanetary magnetic field sets an upper limit on the number of monopoles that crisscross the solar system.

A reliable way of counting monopoles in the laboratory is made possible by the phenomenon of superconductivity, the disappearance of electrical resistance at low temperatures. Again, the principle of detection rests on analogy with electric charge: When a charge moves, it is surrounded by a magnetic field that follows it like a halo. In the same way, a moving monopole is accompanied by a ring-shaped electrical field. If this doughnut happens to encounter a coil of wire, it will accelerate the electrons in the metal and generate a surge of current. If the coil is a superconductor, the current will continue to flow once it is produced. A superconducting monopole detector is therefore easy to build, at least in principle: Keep a coil cold, make sure it has no current in it, and wait. If a monopole from outer space should obligingly pass through it, a current would suddenly appear, and persist. By measuring the magnitude of the current, you can deduce the strength of the monopole, long after the particle itself has disappeared into the laboratory floor.

This is the kind of device Blas Cabrera built at Stanford University, and on the afternoon of St. Valentine's Day, 1982, his labor was rewarded. Suddenly, unannounced, an electrical current appeared in the superconducting coil that had been quiescent before. A monopole had left its unambiguous calling card.

Cabrera's graph of current versus time for that day is a beautiful thing. Regardless of its interpretation, it is a textbook example of graphical excellence. From 3:00 A.M. until after noon, the recording pen holds steady at zero, except for a faint tremor, a sixteenth of an inch high, caused by inevitable electrical noise. Then, at a few minutes before 2:00 P.M., the needle suddenly shoots straight up about an inch, and resumes its trembling at the higher level until midnight. In scientific parlance, you couldn't ask for a cleaner signal. What's more, the strength of the monopole had precisely the value predicted by Dirac.

The reaction of the scientific establishment to Cabrera's paper was keen interest and cautious enthusiasm. Contrary to instances of shoddy science, such as the original experiments on cold fusion, the discovery violated no known principles, the report seemed complete and forthright, the experimental design appeared to be flawless, and the result was unambiguous. Only one thing was lacking for a Nobel Prize, and the immediate rewriting of all textbooks on electromagnetism: independent confirmation. That confirmation never came.

During the following eight years, the experiment was repeated in other laboratories around the world, with negative results. Cabrera and his group improved the sensitivity of the apparatus by a factor of two thousand, principally by increasing the number of superconducting coils and enlarging their size, by detecting the monopole in two places, instead of just one, and by extending the running time to a year and a half. No monopoles showed up. Based on this disappointing result, the authors concluded, in an article published in February, 1990 in the prestigious journal *Physical Review Letters,* that the earlier event was too improbable to be significant and should be "discarded."

The meaning of this recommendation is not obvious. If Cabrera had found an error in his experiment, the retraction would be quickly and cheerfully accepted. Science proceeds by trial and error, and while the penalty for scientific fraud is professional excommunication, there is no opprobrium attached to honest error. When, for example, the recent observation of a neutron star that rotated thousands of times faster than it should have was attributed to a nearby faulty TV camera, astronomers were understanding. But Cabrera did not find any errors in his experiment.

Heinz Pagels, in his book *Perfect Symmetry,* which treats the subject of magnetic monopoles in some detail, calls the St. Valentine's Day event a "fluke." A fluke is a chance event, like a random shot in basketball that happens to score. But no matter how rare a fluke may be, in a rational universe it still has natural causes. Cabrera's observation may have had a silly, accidental cause, like a fly landing on the detector, or it might have been the real thing. The word *fluke* does not differentiate between the two possibilities. Since the distinction is crucial for physics, Pagels's remark does not illuminate the meaning of Cabrera's recommendation.

From the scientific point of view, the retraction does not matter very much. The new experiment, being both more sensitive and more reliable, naturally carries more weight than the older one. Experimentalists will never be able to prove that there are *no* monopoles, but they can set lower and lower limits on the number that hits a square inch of the earth's surface per second, on average. In the estimation of this average, the St. Valentine's Day event will play an increasingly insignificant role, until it will be drowned out by sheer statistics. As the expectation of finding monopoles is reduced by experimentalists, theorists will continue to refine and adjust their models to fit the evidence.

On a more human level, however, Cabrera's retraction represents a loss. As long as that one monopole had a chance of being real, Dirac's pretty mathematics had a chance of being a fundamental law of nature. Furthermore, textbooks on elementary particles and cosmology had a concrete example of how a monopole would make its presence known, rather than just a hypothetical discussion of the subject. If the retraction is heeded, they will in the future have to state flatly that monopoles have never been observed, period. And that makes physics a little less exciting. For science thrives on anomaly, inconsistency, controversy, and doubt. Certainty kills it.

16.

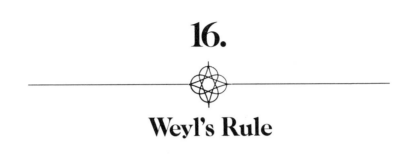

Weyl's Rule

The roundest object on earth is a sphere of niobium-coated quartz, about the size of a Ping-Pong ball. Weeks of painstaking polishing have reduced its irregularities to a depth of less than a millionth of an inch. Supported, without friction, by the electrical repulsion between a charge on its surface and a like charge on its hollow quartz container, the shiny sphere turns silently, rapidly, and, because it has no marks or scratches, imperceptibly. Its axis, like that of a spinning top, does not wobble, not even by so much as a hair's breadth, making the sphere an ideal direction indicator.

Sometime soon, its Stanford University designers hope, this gyroscope will be placed on a satellite in orbit around the earth. The ball's axis will be pointed precisely at the star Rigel, in the constellation Orion, and then, at a pace so slow that only the most sensitive of instruments will be able to detect it, the axis will stray from its guiding star—not because the satellite's orbit has been disturbed or because of the motion of Rigel. Those influences, as well as a host of others, have been taken into account. Rather, the axis will turn because direction itself will have turned. Space, as Albert Einstein surmised more than seventy years ago, is warped in the vicinity of massive planetary objects. The purpose of the Stanford experiment is to measure the distortion caused by earth's gravity.

Few physicists doubt the outcome of the experiment. Like all previous tests, the result will probably confirm Einstein's idea, known formally as the general theory of relativity. According to the theory, it is incorrect to think of space as Euclid did, as a three-dimensional version of ruled graph paper, with a grid set of perpendicular directions. Around great masses, such as the planets, the grid is distorted and directions are skewed. An arrow in orbit around the earth will end up where it started, but it will point in a slightly different direction, turned by an amount that can be calculated from a knowledge of its path and the mass of the planet. Therefore, in the general theory of relativity, geometry assumes the formidable role of describing not only spatial but also dynamic relationships. Space, time, and gravity are fused into a universal synthesis.

With general relativity, Einstein found a way to describe a physical force—gravity—in terms of a mathematical conception: geometry. He began his investigation with experimental facts—primarily the observation that all objects, regardless of weight, fall at the same rate. To this he added general physical principles, such as the proposition that the laws of nature do not change from one day to the next or from one region of the world to another. This combination of empirical evidence and logical requirements led to a geometrical framework that differed from what physicists had been used to.

What was surprising about this new mathematics was that it wasn't new at all. Several decades earlier, mathematicians had invented a similar framework, without suspecting that their construct, called non-Euclidean geometry, would ever be relevant to the physical world. No more did Apollonius of Perga dream, when he computed the formulas for parabolas and ellipses in the third century B.C., that two millennia later Isaac Newton would prove that cannonballs and planets follow just such paths.

How can pure mathematical speculation produce equations that accurately describe the world? No one knows. The "unreasonable effectiveness of mathematics in the natural sciences," as the American theoretical physicist Eugene Wigner called it, is one of the central mysteries of science. But mystery or not, this unreasonable effectiveness has made much of twentieth-century science possible. The success of general relativity, in particular, established the style of theoretical physics, which does not hesitate to reach for even the most imaginative abstractions of pure mathematics to account for experimental observations.

Inspiration can flow in the other direction, too: new mathematical ideas can grow out of the practical requirements of science, as was the case with calculus, which Newton invented to solve problems in mechanics, and even with geometry, which emerged to meet the needs of early builders and surveyors. In this century the reciprocal relationship between mathematics and physics is exemplified by gauge theory, a radical form of geometry that the German mathematician Hermann Weyl proposed shortly after Einstein introduced general relativity.

Inspired by Einstein's conception of gravity, gauge theory led, eventually, to novel insights into the physical world. Indeed, Weyl's creation has become the conceptual scheme most likely to underlie a future theory of everything—a framework for describing all the particles and forces that occur in nature.

At the same time, gauge theory has become a powerful tool at the cutting edge of modern mathematics. The story of gauge theory's troubled birth and metamorphosis illustrates how physics and mathematics run parallel and intersect, often with fruitful results.

The idea of curved space was best illustrated by Einstein himself in *The Evolution of Physics,* written with the Polish physicist Leopold Infeld. "Imagine a town consisting of parallel streets with parallel avenues running perpendicular to them. The distance between the streets and also between the avenues is always the same." They went on to describe a subterranean upheaval that causes a hill to bulge, raising streets, avenues, and houses with it. The space represented by the grid was then curved, or non-Euclidean. In the same way, the earth causes a curvature of space, the effects of which Newton ascribed to gravity.

The first, most sensational test of Einstein's theory, in 1919, measured the deflection of starlight by the sun. Like a car following a street around the side of the hypothetical hill, a beam of light strays slightly from its straight path when it passes a massive object. The aim of the Stanford experiment is to observe the behavior of an antenna sticking straight up from the car's roof. As the car travels over the hill, its antenna will tilt, and it was on the antenna, as it were, that Weyl focused his attention.

Weyl observed that the antenna can be thought of as a vector, a mathematical construct, which, like an arrow, is defined by its magnitude (or length) and its orientation in space (or direction). Weyl argued that if, as general relativity holds, the direction of a vector is subject to perturbation, so should its length. If, under the influence of gravity, space warps, distances should shrink and stretch as well, just as they do when the image of a square grid is projected onto a warped screen. In other words, the antenna's length should depend on its location to the same degree that its direction does.

From a mathematical point of view, Weyl's proposal was convincing, for nothing in geometry suggests that the direction of a vector is more or less fundamental than its magnitude. To mathematicians, the two variables are perfectly equivalent.

Weyl named his new geometry gauge theory, after the gauge blocks that machinists use to measure lengths. And he found, to his delight, that the equations that govern the variations of distance scales were identical with the equations of electromagnetism. Because of this correspondence, Weyl concluded that just as gravity bends space by tilting directions, electricity and magnetism warp space by distorting its distance scales. Thus, Weyl seemed to have found a way of incorporating general relativity and, therefore, gravity, along with electricity and magnetism, into a single, all-encompassing theory.

Certain that he had made a major discovery, Weyl sent Einstein a paper describing his theory. The response was swift and devastating. On a postcard dated April 15, 1918, Einstein acknowledged the beauty of Weyl's conception even as he denied its physical significance. He sketched the rudiments of an experiment like the one in preparation at Stanford to demonstrate that predictions based on Weyl's theory would contradict some of the most fundamental assumptions about the material world. Illustrated with a minuscule diagram, Einstein's objection amounted to showing that two atoms taken on two different journeys through space, and thereby exposed to different degrees of electromagnetism, would, by Weyl's reasoning, differ in size when brought back together again. Similarly, Einstein argued, the length of a common ruler would depend on its history—on where it had been. "I must confess," he wrote, "that in my opinion it is impossible that your theory corresponds to nature."

Einstein was not insensitive to the aesthetic dimension of gauge theory, which, to Weyl, constituted the strongest argument in its favor; even his postcard referred to the idea as a stroke of genius. But the theory's formal beauty was not enough to outweigh the lack of evidence for its underlying assumptions and the profound ways in which it contradicted what was known about the world. The difference between length and direction may be insignificant in geometry, but in physics it is crucial.

With the advent of quantum mechanics, some seven years later, Weyl saw that gauge theory could be recast in an unexpected way. The key to this reinterpretation is the treatment in quantum mechanics of a subatomic particle as a wave. The wave's amplitude, or its height, represents the probable location of the particle with which the wave is associated: where the amplitude is greatest, the particle is most likely to be found. The speed of the wave stands for the particle's speed, and the direction of the wave's motion, the direction of the particle's motion.

The only attribute of a quantum mechanical wave that lacks a counterpart in the usual description of a particle is the phase, the variable that denotes the fraction of a cycle from peak to trough, and is used to quantify cyclical phenomena. Consider, for the sake of illustration, two clocks, one whose second hand points to the number three, the other whose second hand points to six. The hands are out of phase by one quarter of a cycle. Moreover, this distinction means something: the hands represent two different times of day. In quantum mechanics, the phase seemed to have no observable significance; a particle described by one wave could also be described by the wave's inverse, with which it was completely out of phase.

Because predictions based on wave theory have proved remarkably accurate, this peculiarity in the quantum mechanical model of particles didn't concern physicists. But Weyl, the mathematician, was bothered. Surely, he reasoned, if all the other attributes of a wave contribute to a particle's description, the phase should too. He speculated that a particle's phase varies according to where the particle is located in space, just as the distance scale had varied in his earlier theory. When he derived the equations that the phase has to obey to make this speculation consistent, he found they are identical with the equations that describe electromagnetism—just as his earlier gauge equations had been. This time, however, there was no experimental evidence to contradict Weyl's conclusion: quantum mechanics, by introducing the seemingly irrelevant phase variable, *required,* and thus *explained,* the existence of electromagnetic forces.

No analogy can adequately explain gauge theory, but the effort might help to convey its significance. Consider a baize cloth stretched out on the ground, with a blue bowling ball lying at its center, and another, tiny object, visible only as a white dot, somewhere near an edge. High above, a balloonist is watching. When he notices that the dot has begun to move toward the bowling ball, he sets himself the task of explaining the phenomenon.

It first occurs to the balloonist that the bowling ball may be exerting an attractive force on the small white object. This is an analogy for the notion of action at a distance, which underlies Newton's conception of gravity and was later applied to both electricity and magnetism. Although useful, the notion that one object can pull another across empty space without the benefit of an intermediary is, in Newton's words, "so great an Absurdity, that I believe no man who has in philosophical Matters a competent Faculty of thinking, can ever fall into it."

When the balloonist grows similarly disenchanted with that explanation, he surmises that the cloth is suspended a few inches above the ground, sagging a bit under the bowling ball's weight. In that case, the white object is sliding downhill toward the middle, an outlook consistent with general relativity. The baize cloth represents curved space, and the white object is affected only by the steepness of the slope in its immediate vicinity. This is the way a modern physicist conceives of gravity.

But suppose the force at work is not gravity. Now the balloonist proposes still another explanation, one that stands as an analogy for gauge theory. Imagine, in this instance, that the cloth is lying flat again; space is not curved. The balloonist decides that the white object is in fact a Ping-Pong ball, and assumes that it is so smooth that any rotation it made would be undetectable, even in principle. And if the white object was indeed rotating, it would roll toward the bowling ball. This image suggests how a modern physicist thinks of electromagnetism as it operates upon subatomic particles.

In the actual quantum theory, electromagnetic forces have nothing to do with the turning of elementary particles in ordinary three-dimensional space. The idea of rotation is merely a way of suggesting the cyclical, unobservable nature of the phase variable. This rotation takes place in a hypothetical framework called internal space. Theoretically, the unobservable axes of another space may have one, two, three, or any number of dimensions extend from each point in real space. These internal spaces are connected to one another and guide the motion of a particle along its orbit in ways that match the observed effects of electromagnetism.

The ways in which general relativity and gauge theory were developed exemplify the differences between the methods of the physicist and those of the mathematician. Einstein had been compelled, somewhat reluctantly, to accept a strange brand of mathematics—non-Euclidean geometry—to account for empirical observations; he restricted speculation to variables about which he had no experimental information. In contrast, Weyl, using an argument whose sole strength was mathematical consistency, speculated about things—namely, distance and size—that scientists had been observing for some time. His error lay in ignoring the constraints of the already known. The manner in which Weyl revised gauge theory—recasting it in terms of quantum mechanical phases—came closer to Einstein's way of doing physics.

Weyl wasn't the only one to undergo a shift in perspective. By the 1950s, when he and Einstein were working side by side at the Institute for Advanced Study, in Princeton, Einstein had come to feel that the only way to overcome the virtual lack of experimental evidence for the relationship between gravity and electromagnetism was to make bold mathematical guesses. Weyl, on the other hand, had learned his lesson, and put more trust in the experimental evidence than in pure mathematical speculation. The physicist and the mathematician had exchanged roles.

Meanwhile, gauge theory had taken on a life of its own. Inspired by Weyl's investigations, the French mathematician Élie Cartan in the 1930s laid the foundation for fiber bundle theory—the mathematical description of internal spaces. Interested purely in elaborating on the formal aspects of gauge theory, he discovered the rules for how the axes of neighboring internal spaces must be connected to one another so that the whole edifice doesn't get tangled up. Fiber bundle theory has since become a useful mathematical tool.

On the other side of the fence, in physics, gauge theory was modified in such a way that it could describe forces other than electromagnetism. In 1954, Chen Ning Yang and Robert Mills, both at the Brookhaven National Laboratory, on Long Island, used gauge theory to account for the behavior of the strong nuclear force, which holds together the atomic nucleus. They concocted a more complicated internal space than Weyl had. The equations describing this space now underlie the modern view of all particles and forces (except gravity) and have been used to predict the existence of particles to carry those forces—particles that have since been found in the debris of atomic collisions in high-energy accelerators. Gauge theory, Weyl's ugly duckling, has turned into a swan of surpassing beauty and power.

Cartan, driven by an interest in the theory's formal elegance, and Yang and Mills, striving to find a description of the strong nuclear force that would unite it with that of electromagnetism, could scarcely have communicated, so different were their concerns. But occasionally, the elaborations of gauge theory in mathematics and in physics have crossed paths. In 1986, Simon Donaldson, a young Oxford professor, won the Fields Medal, the equivalent, in mathematics, of the Nobel Prize, for his application of the Yang-Mills equations to the investigation of four-dimensional spaces—a problem with no apparent connection to physics.

In the branch of pure mathematics called topology, the concept of dimension is unencumbered by references to the physical world. When topologists speak of a five-dimensional space, for instance, it takes five parameters, or numbers, to describe any point in that space. Such imaginary spaces can have as few as one dimension and as many as an infinite number of them. Besides investigating the properties of the spaces themselves, topologists try to identify properties that different spaces have in common, an effort that to mathematicians is the equivalent of a biologist's developing a taxonomic scheme for organisms.

The usual tack in exploring the nature of mathematical spaces is to compare topological properties—for example, the way in which two-dimensional spaces embedded in four-dimensional spaces intersect. Donaldson took a different approach. Noting that the Yang-Mills equations of physics refer to a four-dimensional space, he began with the equations and reasoned backward to the implications for topological spaces. He discovered that four-dimensional spaces are unique. In a space of any other dimension, the rules that connect neighboring points can be altered slightly without affecting the global properties of the space. In four dimensions, on the other hand, the rules seem extraordinarily sensitive; change them but a little, and the entire space becomes radically rearranged.

Donaldson's unusual approach solved a substantial problem in the classification of multidimensional spaces. But more intriguing was his discovery that the imaginary space that happens to correspond to the real world is fundamentally different from all other spaces. These findings, and the unconventional way in which they were made, stunned the mathematics community.

One man who would not have been shocked was Weyl. Throughout his career, he wrote mathematics books for physicists and physics books for mathematicians, assuming the role of messenger between the two disciplines and, in so doing, enhancing them both. He was, in fact, a living embodiment of the rich, though largely enigmatic, relationship between mathematics and physics.

17.

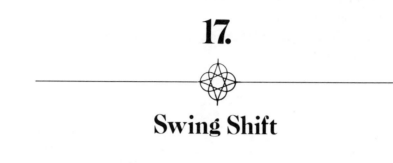

Swing Shift

Jacopo Belbo, the hero of Umberto Eco's novel *Foucault's Pendulum,* is consumed by a common human yearning—a desperate thirst for the absolute. He finds solace in contemplating the motion of Foucault's pendulum, an instrument displayed in science museums throughout the world to demonstrate the daily rotation of the earth. A heavy sphere, usually made of brass, is suspended from a long wire and set into oscillation. A concealed magnetic mechanism delivers gentle pushes to keep the pendulum swinging in spite of air resistance.

The swing of the wire and bob defines a single vertical plane, but that plane itself moves almost imperceptibly. It rotates with respect to the floor, turning by a full 360 degrees in a period that depends on its earthbound position. In Paris, where Belbo watches it, the circuit takes about thirty-two hours. Following popular accounts, Eco explains that the floor, and indeed the whole earth, rotates beneath the pendulum, "whose own plane never changed direction."

According to this description the plane of Foucault's pendulum is fixed in space, attached by some intangible influence to an immobile framework that Newton called absolute space. For a disillusioned modern man such as Belbo, watching the brass sphere as it follows the inexorable dictates of a remote firmament is a wondrous hint of transcendent certainty—a way, he says, of finding God again.

Actually Eco's explanation is not quite right. The plane of a pendulum suspended over the North Pole does remain fixed with respect to the stars. The rotation of the pendulum as seen by someone whose feet are firmly planted on the polar ice corresponds directly to the earth's own daily rotation. But a pendulum's rotation slows down with increasing distance from the pole until it stops altogether at the equator. A Foucault pendulum in Singapore, which virtually straddles the equator, furnishes a particularly clear example of the flaw in Eco's reasoning.

If the Singapore pendulum is set swinging east and west, parallel to the equator, its plane will indeed remain fixed with respect to both the earth and the stars as the earth turns. But if the swing is perpendicular to the equator, the pendulum's plane will sweep through space as the earth rotates, even though the swing will again appear fixed to a ground-based observer. (If you hold your hand parallel to the ground as you turn around, it will stay in the same plane, but if you hold it perpendicular to the floor, as if to shake hands, its plane will turn with you.) Anywhere except at the North and South Poles, in fact, the plane of Foucault's pendulum turns with respect to the stars. As a result, it makes something less than a complete circuit in twenty-four hours.

Nevertheless, Eco's metaphor is apt in a more modest sense. The motion of the brass sphere represents a subtle compromise between the external forces acting on it—gravity and the tension of the wire as it resists gravity—and the innate tendency, called inertia, of any body to preserve its direction of motion with respect to the stars. The external forces change direction continuously as the earth rotates, but the inertia of an object is an intrinsic quality, independent of the external environment and the forces acting on the body. In some sense it deserves the adjective *absolute*.

Since the invention of Foucault's pendulum in 1851, its motion has been understood in terms of gravity, tension, and inertia. Recently, though, a new analysis opened a fresh perspective on the old theory—a perspective very different from Belbo's. In this view the angle through which the plane of the pendulum rotates as it is carried around the earth at any latitude is a simple consequence of the geometry of the earth-pendulum system. It is but one instance of a general phenomenon that arises in many other systems with similar geometric properties when they make a round-trip of some kind. Such changes in phase, or angle, crop up both in the microscopic world of quantum mechanics (where many other familiar laws are suspended) and in the macroscopic world of pendulums. Viewed as an instance of this phase shift—technically known as the geometric phase, or Berry's phase, after the University of Bristol physicist Michael Berry, who drew attention to it in 1984—the rotation of the pendulum speaks not so much of the absolute as of the universal.

Although Berry discovered the geometric phase in the domain of quantum mechanics, he relied on an example from the familiar, macroscopic world to illustrate the concept. Imagine a globe with a short pencil lying flat on its surface at the North Pole and pointing toward London. Now slide the pencil southward along the meridian, through London, all the way to the equator. Then, without turning the pencil, slide it sideways, following the equator eastward to Singapore, at 103 degrees east longitude, and finally move it eraser-first up the Singapore meridian and back to the North Pole.

In the end the pencil comes to rest where it was before, but it has turned counterclockwise through an angle of 103 degrees—even though throughout its trip it has pointed in the same direction, due south. The change in direction during the round-trip is a macroscopic example of Berry's phase. The phase shift takes place even though the pencil is not rotated, and its magnitude is not affected by the speed of the pencil as it makes the loop. The shift simply reflects the fact that the loop is completed, and its magnitude depends only on two quantities: the shape of the path of the pencil (the direction change would have been larger if the pencil had followed the equator eastward to the mid-Pacific) and the geometry of the space in which the path is embedded (the globe's spherical surface).

The phase shift of the pendulum is not quite so easy to visualize. But it too depends on the fact that the pendulum travels in a closed loop around a sphere, because of the earth's rotation. Here the loop is a circle of latitude instead of a spherical triangle. No person or mechanism adjusts the plane of the swing as the loop is made, but at the end of the day (except at the poles and the equator) the pendulum does not return to its original swing plane, as one might expect. Instead it is displaced by a certain angle—by 270 degrees for a pendulum in Paris.

As the pencil and the pendulum each suggest, there can be no change of angle unless the object in question has a direction of its own. A mathematical point or a more symmetric object, such as a ball, would not exhibit the shift. Imagine a clear glass marble following the path traced by the pencil; no change would be visible when it returned to its starting point. Curiously, in quantum mechanics even a pointlike particle has a kind of direction, and so the possibility is always lurking that a geometric phase will develop. Perhaps this helps explain why Berry's phase was discovered in the arcane arena of quantum theory before it was identified in the everyday world.

In quantum theory the basic scheme for describing a physical system—be it an atom, a nucleus, an elementary particle, or a collection of particles—is a wave function. A wave function is a mathematical construction that, rather like a blueprint of a machine part, encodes a complete description of the system. Those who can read it can thereby make predictions about how the system will act. For a particle, the wave function describes real, measurable attributes, such as position and momentum. It also describes an attribute that cannot be measured directly: the phase of the wave function. That phase can be understood through analogy with more familiar waves.

For an ordinary wave—a wave on a lake or in a beam of light—phase is simply a number that describes the stage of the wave in its oscillation cycle. The phase is expressed as an angle: 360 degrees is a full cycle, 180 degrees a half-cycle, and so forth. Two waves are said to be in phase when trough matches trough and crest matches crest; when one wave is shifted with respect to the other, the phase angle measures the shift. For example, two otherwise identical waves that align crest to trough and trough to crest are said to be 180 degrees out of phase.

A difference in phase can be observed as an interference pattern, which forms as the two waves reinforce or cancel out each other. Indeed, for quantum-mechanical wave functions, phase can be detected only through interference patterns.

Thus phase can generally be observed only when particles interact.

The physical significance of the quantum-mechanical phase has been understood since the birth of quantum mechanics in 1926. But because it is often without measurable consequences, physicists tended to be less careful about calculating it than other features of the wave function, such as its magnitude or direction. Even when they did turn their attention to the phase, they often introduced certain simplifying assumptions to make the calculations more convenient. The result was that they missed a part of the phase altogether—the part Berry calls the geometric phase. In terms of the pencil on the globe, they accounted for the external forces acting on the pencil to turn it or keep it straight, but they missed the rotation automatically introduced by a round-trip on the globe.

In the quantum-mechanical case the globe need not be a real surface in three-dimensional space. The excursion of a particle can take place on a surface in an abstract space called parameter space, in which the directions up, down, and sideways represent quantities that affect the particle or its environment. Parameter space can have many kinds of dimensions. The axes of the space can mark off the length, width, or height of ordinary space; they can also indicate temperature, say, or the momentum of the particle, the strength of an external electric field, or the density of the surrounding medium.

It was no surprise that a change in any of these quantities—a one-way trip through parameter space—could affect the phase of a wave function. What was a surprise, and something of an embarrassment to physicists, was Berry's insight that a closed circuit in parameter space—a round-trip in physical space, perhaps, or a cycle of heating and cooling—can also have an effect on phase.

Berry is a quantum mechanic by trade, so it is not accidental that one of the first applications of his discovery was to the interior of a molecule. He was engaged in a theoretical study of the simplest of all molecules, the charged hydrogen molecule, composed of two protons held together by a single electron. He calculated the wave function of the electron moving in the electric field generated by the protons. Next he asked what would happen if the protons slowly moved apart, reshaping the electric field, then converging again—in effect, taking the electron on a round-trip through parameter space. It had been assumed that after such an excursion, which after all restores the environment of the electron to its initial state, the electron's wave function would also return to its initial state. But by scrupulously following the rules for calculating wave functions, Berry showed that the phase would have suffered an unexpected rotation—the geometric phase.

For Berry, this rather specialized finding smacked of the universal. He suspected that the geometric phase would arise in many other circumstances. Since the phase of a wave function is usually more difficult to measure than its magnitude, he searched for special conditions in which the geometric phase would be easier to detect. He found the solution in atomic systems analogous to pencils on a globe: objects endowed with intrinsic direction indicators.

Among such objects are the many kinds of atoms, nuclei, and elementary particles that resemble miniature magnets, and so possess a direction analogous to the north–south polar axis of a magnet. Photons (the particles of light) also assume a preferred direction when their oscillations are polarized, or oriented in a specific direction, by Polaroid sunglasses, for example. In his first article on the geometric phase, Berry predicted that when such particles are taken on a round-trip through parameter space, they acquire a measurable geometric phase. For such particles parameter space is sometimes just our ordinary space, so the pencil metaphor illustrates their motion directly.

In the years following Berry's prediction physicists around the world took up the challenge of testing his idea. The geometric phase seemed to arise in any number of settings, but a minor controversy soon ensued. In one of the first tests, polarized light was channeled through an optical fiber. When the fiber was coiled, so that the photons did a loop-the-loop, their polarization direction turned through an angle corresponding to the geometric phase, as Berry's quantum-mechanical analysis predicted. Yet the angle can also be explained in terms of Maxwell's equations of electromagnetism, which describe light without reference to quantum mechanics. A debate arose about whether Berry's phase can properly be called quantum-mechanical.

In short order, however, other tests decided the question. Nuclei of chlorine atoms and even neutrons—purely quantum-mechanical systems that must be described by wave functions—exhibited a geometric phase when they were rotated by magnetic fields or (for the chlorine nuclei) when a crystal partly made up of chlorine was simply turned. Since the phase shift clearly operates in the quantum world, calling it quantum-mechanical or classical in the experiment with polarized light turned out to be a matter of preference.

At the quantum level the phenomenon is certainly, in Berry's words, "remarkable and rather mysterious." One mystery lies in the formula Berry derived for the geometric phase, which expresses it as a function of the *area* in a parameter space enclosed by the round-trip. For the pencil the relevant area is that of the great spherical triangle bounded by the London meridian, the equator, and the Singapore meridian. Yet the pencil does not sweep out this entire area, but only traces out a path around its periphery. How can the oriented object "know" the overall shape of parameter space when it visits only its edge? For a macroscopic system, such as the pencil and globe, the answer can be found in elementary geometry. At the quantum level, however, the mystery is deeper. In the end one has to ascribe the enigma to the peculiar ability of quantum-mechanical particles to be spread out in space, rather than localized in a point.

In the macroscopic realm each instance of Berry's phase can be explained in an alternative, usually more cumbersome way through the familiar laws of classical physics. Just as the turn of the pencil is a consequence of elementary solid geometry, the rotation of the polarization of light follows from Maxwell's equations, and the motion of Foucault's pendulum can be derived from Newton's laws. What Berry's analysis did was supply a single, simple way of thinking about all such disparate phenomena. Even if it is not mysterious, the classical version of Berry's phase is just as remarkable as its quantum-mechanical analog.

Consider the ability of Berry's analysis to link two phenomena that could hardly seem less alike: the rotation of a pencil on a globe and the uncanny ability of a falling cat to land on its feet. The cat's behavior is intriguing enough to have led countless curious boys to drop the family cat off the garage roof. After all, a rigid body, such as a brick, dropped without any spin cannot change its orientation; the law of conservation of angular momentum (which follows from Newton's laws) declares that an object cannot acquire rotation unless an outside force acts on it. Why should a cat be different?

To sharpen the puzzle, one should note that a cat can right

itself even when it is dropped carefully, without any hint of an initial twist. What is more, it does so not by rowing through the air, so to speak, as it might underwater. Air is far too tenuous for that strategy to work in the short time of the fall. In the latter half of the nineteenth century, when classical mechanics was already well established, physicists were so skeptical about the cat's rotation that they, like schoolboys, felt obliged to perform the experiment.

Accordingly, reluctant cats were dropped upside down and photographed. The pictures were published in scientific journals, and quaint and grainy as they appear, they are unambiguous. The cats began to fall upside down, twisted their bodies in a couple of supple moves, and hit the floor paws first. Had feline cunning outwitted Newton's genius?

Analysis eventually revealed the truth, but it is so startling that even professional physicists doubt it when they first hear the explanation. In fact, nonrigid bodies *can* change orientation without the influence of external forces. Newton's laws are saved by the fact that even though the cat twists parts of its body as it falls, it is twisting other body parts in the opposite direction, preserving a net angular momentum of zero. How, then, can the cat achieve a net rotation?

It does so by a technique reminiscent of a figure skater's. A twirling skater can slow down and speed up again simply by first extending her arms and then drawing them in. In the process she takes a round-trip in the parameter space that describes the position of her hands, during which time her orientation shifts in a special sense.

The shift can be seen by imagining a second skater. The two skaters start out spinning precisely in sync. After making an excursion in parameter space, the first skater will again be rotating at the same speed as the second one, but an observer will no longer see both skaters' faces at the same time. In slowing down and then speeding up, the first skater has gained a geometric phase.

The cat incorporates this trick into a larger pattern of movement. As it falls it draws in its legs and rotates its body in one direction while twirling its tail in the opposite direction (thereby conserving angular momentum). Then it stretches its legs and reverses the process. The outstretched legs keep the cat's body from twisting all the way back to its original orientation. Like the skater, the cat traverses a closed circuit in parameter space. As a result, it develops a geometric phase of 180 degrees and lands safely.

Thus falling cats join Foucault's pendulum, looping photons, and vibrating molecules in the gamut of phenomena Berry has brought together into one coherent account. How a cat rights itself had been understood before, but Berry's insight cast the explanation in a new light. Having found a minor error in a quantum-mechanical calculation, he not only corrected it but proceeded to track down its mathematical essence, and in so doing identified a profound relationship that inspired others to apply a similar analysis to a multitude of cases.

Even though Michael Berry and Jacopo Belbo of Eco's novel each sought to probe beyond appearances, they approached the problem of Foucault's pendulum from opposite directions. Belbo, in his mystical quest for the absolute, found traces of it in the particulars of this world. Berry, the scientist trying to understand the particulars, found evidence of the underlying unity and simplicity of nature, and thus discovered the universal.

About the Author

HANS CHRISTIAN VON BAEYER is the author of *Taming the Atom*. His essays about the meaning of physics in numerous publications, including *Discover*, *The Sciences*, and *Reader's Digest*, have won him such honors as the Science Journalism Award of the American Association for the Advancement of Science and the National Magazine Award. He is a professor of physics at the College of William and Mary in Williamsburg, Virginia, where he lives with his wife and their two daughters.

About the Type

This book was set in Electra, a typeface designed for Linotype by W. A. Dwiggins, the renowned type designer (1880–1956). Electra is a fluid typeface, avoiding the contrasts of thick and thin strokes that are prevalent in most modern typefaces.